Social and Environmental Impacts of the
James Bay Hydroelectric Project

Planning and construction of the James Bay Hydroelectric project began in the early 1970s, when the effect of such projects on the physical and social environment was seldom considered. As the project matured, however, its unique and diverse environmental impacts came under intense scrutiny on both sides of the border.

The first mega-scale hydro project to be built in the sub-Arctic, capable of generating as much electricity as fifteen nuclear power plants, its impact includes disruption of vast areas in an extremely fragile ecosystem as well as displacement of native peoples and the introduction of dangerous levels of mercury into their food supply. The debate over these complex environmental issues has been further complicated by political issues stemming from the importance of the project to the economic development of Quebec and the sale of at least ten per cent of the electricity generated to the United States.

The contributors examine core issues of the controversy both in relation to James Bay and to other large hydroelectric projects, such as the Aswan dam in Egypt and the Three Gorges dam in China. Providing insights from an unusual variety of disciplines, the authors offer important considerations that must be taken into account as Quebec assesses additional phases of hydroelectric development of the watershed east of Hudson Bay.

JAMES F. HORNIG is professor emeritus of chemistry and environmental studies, Dartmouth College.

McGill-Queen's Native and Northern Series
Bruce G. Trigger, Editor

Social and Environmental Impacts of the James Bay Hydroelectric Project

Edited by

JAMES F. HORNIG

McGill-Queen's University Press
Montreal & Kingston · London · Ithaca

© McGill-Queen's University Press 1999
ISBN 0-7735-1836-3 (cloth)
ISBN 0-7735-1837-1 (paper)

Legal deposit fourth quarter 1999
Bibliothèque nationale du Québec

Printed in Canada on acid-free paper

McGill-Queen's University Press acknowledges the
financial support of the Government of Canada through
the Book Publishing Industry Development Program
(BPIDP) for its activities. We also acknowledge the
support of the Canada Council for the Arts for our
publishing program.

Canadian Cataloguing in Publication Data

Main entry under title:
Social and environmental impacts of the James Bay
Hydroelectric Project
(McGill-Queen's native and northern series ; 18)

Includes bibliographical references and index.
ISBN 0-7735-1836-3 (bound) – ISBN 0-7735-1837-1 (pbk.)

  1. James Bay Hydroelectric Project–Environmental
aspects. 2. James Bay Hydroelectric Project–Social aspects.
3. Cree Indians–James Bay Region (Ont. and Quebec)–Social
conditions. 4. Environmental impact analysis–James Bay
Region (Ont. and Quebec) 5. James Bay Region (Ont. and
Quebec)–Social conditions.   I. Hornig, James F., 1929–

HD9685.C33J35 1999   333.91'A'09714115   C99-900826-9

Typeset in 10/12 Baskerville by True to Type

Contents

Tables and Figures

Preface

The James Bay hydroelectric development project constitutes an unusually rich case history of the physical and social environmental impacts accompanying a modern-day megaproject of economic development. Despite its identification with James Bay, the components of the proposed Hydro-Québec development project actually range along all the watersheds on the east shores of Hudson Bay from the southern tip of James Bay to Ungava Bay on the north, a distance of 1,200 kilometres. Although only one of the three originally planned components of the James Bay project has been completed as of 1998 and much of the remainder postponed indefinitely, the largely complete La Grande River phase of the project already has an electrical generating capacity of over ten thousand megawatts (roughly the equivalent of ten large nuclear reactors), and the impoundment reservoirs have flooded a land area of more than ten thousand square kilometres – nearly as large as the state of Connecticut.

Launched by Premier Robert Bourassa in 1971, the project was a central element in Quebec's plans for economic development. At that time neither physical nor social environmental impact analyses were of much concern but it quickly became clear that every conceivable dimension of the ecological, social, ethical, and political environmental agenda would be raised in connection with the James Bay project.

The native people of the area, the Cree and the Inuit, were the first to challenge the project, and their complaints found many supporters in Canada and across the border in the United States. When the vast scope of the project became better known to scientists across North

America, concerns about its impact on the flora and fauna of the fragile sub-Arctic ecosystem were raised. As the debate about the physical and social environmental impacts spread, a number of national governance issues also came to the fore, many of which were exacerbated by the ongoing activities surrounding Quebec's separatist initiatives.

With roughly ten percent of the power scheduled for export to the United States, the debate quickly spilled into a North American context. Indeed, us environmental groups proved at least as vocal as their Canadian counterparts in questioning the project on every score. Frequently joining with the Quebec Cree, who opposed the development, the us groups not only alleged a violation of the rights of native peoples and a disregard for the ecological impacts of the project on wildlife and natural systems but they also challenged the morality of importing such environmentally tainted power into the us. Nevertheless, the environmental position that eastern North America's energy needs would be better served by conservation, or at least by renewable energy sources, was complicated by the fact that hydroelectricity has certainly been more popular with the environmental community than fossil fuel or nuclear-powered plants.

The belated recognition in Canada of the need for an environmental impact assessment introduced new complications, bringing into uncomfortable focus questions regarding the relation between provincial and national authority in such matters, while at the same time raising the stakes regarding the growing sensitivity in Canada to the interests of indigenous peoples.

Conflict generated by the divergent values and interests of the various parties was exacerbated by lack of information and, sometimes, by failure to use existing information. The world has had relatively little experience in confronting the ecological impacts of hydroelectric development in extreme northern latitudes, and those experiences have been poorly studied and sparsely documented. For example, the appearance of elevated concentrations of mercury in fish taken from the newly flooded reservoirs seemed to come as a great surprise to most of the actors, despite the (admittedly sparse) literature that might have alerted them beforehand to the problem. Similarly, the death of large numbers of migrating caribou in a river swollen by a man-made flood led to wide public dismay, though scientists were limited in their ability to assign cause and effect, or to put the incident in the context of natural processes.

The authors of this slim volume have attempted to stand back and examine just a few of these issues from the perspectives and principles of a variety of disciplines. Their purpose is to inform and stimulate thoughtful consideration by providing an overall perspective that might serve to broaden the context within which specific issues can be debated. No clear advocacy position emerges. One author is from Newfoundland, the other seven from the United States, but what this sacrifices in terms of continuous local familiarity with Quebec is perhaps more than compensated by somewhat greater objectivity. All of the authors have substantial experience in the Far North, and most have spent time at the James Bay site. While they write from the perspectives of their own disciplines, they do not hesitate to connect their observations to related areas. The results are sometimes surprising.

Oran Young begins with a description of the social issues as seen from the perspective of a political scientist. He discusses governance, rights, and authority on scales ranging from regional interest groups, such as the Cree and the Inuit, to the trans-boundary issues of international affairs. The overview is then continued from the perspective of a team comprising an economist (Warner) and an ecologist (Coppinger). They provide a detailed sketch of the evolution of the James Bay project as a Quebec provincial initiative, and also as a project with major specific impacts on the Cree. In particular, they describe the James Bay and Northern Quebec Agreement of 1975, which has proven so pivotal in discussions of justice and equity in relation to the native peoples.

Coppinger then joins with Ryan, also an ecologist, to look specifically at the ecological impact of dams and reservoirs in this sub-Arctic environment. Proceeding from a geological and ecological characterization of the James Bay region, Coppinger and Ryan use fundamental principles of ecology to make predictions regarding the effects of the project's dams and reservoirs on the biological productivity as well as the survival and diversity of aquatic and terrestrial animal species. Finally, they compare their predictions with experience accumulating from the project.

Roebuck, a toxicologist, examines the methyl mercury problem in detail. What emerges is a fascinating picture of a belated response by Hydro-Québec whereby they informed the Cree after the fact about the hazards of methyl mercury in fish and instructed them regarding mitigative steps, only to trigger what may prove to have been an overreaction with its own unfortunate consequences.

Aspects of the social impact of the James Bay project on the Cree are examined from three very different perspectives by an economist (Warner), an anthropologist (Tanner), and an artist (Woodward). In a far-ranging essay, Warner speaks as a broadly educated social scientist with substantial personal experience of all aspects of the James Bay project. After comparing the current pressures for cultural change with the earlier accommodation of the Cree fur trade culture, Warner suggests that the dams and reservoirs of the hydroelectric project have had less effect on Cree culture than the broadly accelerated impact of western material culture it has brought about, particularly the construction of roads that now connect the region to Montreal and Quebec City.

Tanner, a widely published expert on Cree culture, shows that the non-Western world-view of the Cree can help us understand the social and ecological issues raised by James Bay. He explores the changing reactions of the Cree to the project, from initial resistance to an apparent acceptance through the signing of the James Bay and Northern Quebec Agreement, to their current strong opposition to further development.

Woodward, himself an Alaskan artist familiar with indigenous art, describes contemporary Cree art in its historical context. His sensitive observations as he examines artefacts and objects produced for sale and talks with their creators provide an important insight into the nature of the people who are most affected by the hydroelectric development project. He draws particular attention to the astounding, but almost unrecognized, artistic productivity of this small population and speculates on the links between James Bay and the changes he finds not only in subject matter but in the materials and function of artistic expression among the Cree.

This collection of essays was conceived at a time when the La Grande River portion of the project was substantially finished and the beginning of the next major phase on the Great Whale River appeared imminent. The debate was intense, and as is the case when decisions are made under pressure, it tended to focus on details, sometimes becoming shrill at the expense of the larger principles that might lead to informed decision. Since then the Great Whale project has been postponed. Perhaps these essays, now in the context of the Great Whale postponement, will contribute to a more productive discussion of the merits of continuing with subsequent phases of the James Bay project, and to a reasoned analysis of other megascale development projects.

James F. Hornig
Dartmouth College
Hanover, New Hampshire

Acknowledgments

The authors express their gratitude to the many people who helped educate us about this enormously complex situation. They tutored us patiently and, we hope, effectively to appreciate the many sides of a fascinating and important story: Helen Atkinson, Philip Awashish, Jane Blacksmith, Charlie Brien, Richard Charland, Joan Coffey-Dietrich, Robert Connelly, Claude Demers, Robert Denis, G. Jean Doucet, Marie-Claude Durand, Jacques Guevremont, Michel Hannoyer, Larry House, Kurt Jensen, Guy Levesque, Matthew Mukash, Violet Pachanos, Alan Penn, Shiley Hendricks Perry, Robin Philpot, Robert Schetange, Robert Schwartzwald, Colin Scott, Pierre Senécal, William Sheperdson, Richard Verdon, Leonard Watkins

The authors express particular appreciation to our editor, James Hornig, at Dartmouth College, for his care and skill in coordinating and shaping this complex project across geographic and disciplinary boundaries.

We are grateful to the Government of Canada, the government of the Provice of Quebec, and the Institute on Canada and the United States in the Dickey Center at Dartmouth College for supporting this work.

Comparative data for selected hydroelectric generating sources in North America

Location	Basin area (square kilometres)	Population	Generating Capacity (megawatts)
Missouri River	1,323,700	9,805,800	1,800
Colorado River	827,500	4,442,700	4,500
Columbia River	665,700	5,113,300	21,300
Tennessee River	102,100	3,739,300	1,530
La Grande (phases 1 and 2)	175,300	2,900	15,700
Great Whale (proposed)	64,700	2,200	3,200
NBR (proposed)	129,500	25,300	8,400
	•		
TOTAL AREAS			
United States	9,166,850		
New England plus New York State	309,270		
Province of Quebec	1,540,680		
James Bay region	350,000		
James Bay and Northern Quebec Agreement (JBNQA)	1,066,000		

(Sources: Various Hydro-Québec publications, 1991–92)

1   Major hydroelectric installations of North America (Grande Baleine and Nottaway-Broadback-Rupert [NBR] are projected).

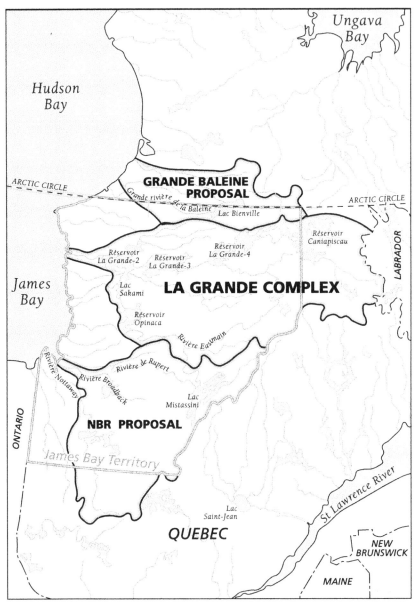

2   The James Bay hydroelectric project: La Grande is completed, Grande Baleine and
    Nottaway-Broadback-Rupert [NBR] are projected.

3   La Grande project, showing reservoirs, generating stations, roads, and powerlines.

PART ONE

Overview

1 Introduction to the Issues

ORAN R. YOUNG

Not so long ago, policy analysts and members of the general public alike equated progress with megaprojects involving the exploitation of natural resources and, more particularly, large-scale efforts to harness the energy latent in the flow of major rivers. On this account, the Tennessee Valley Authority transformed a relatively backward area of the American South into an economically productive region. The Bonneville Power Administration contributed greatly to the industrialization of the Pacific Northwest. The development of the Colorado River played a key role in the emergence of California as a showcase for American agriculture. These American initiatives – largely products of the New Deal and the effort to break the grip of the Great Depression of the 1930s – have had an immense impact on policy making in the rest of the world ever since. Whether we look to China, the Indian subcontinent, Africa, or South America, initiatives that imitate the American model of regional development based on the exploitation of water power on a large scale are much in evidence today.

In recent years, however, these megaprojects – and the assumptions about economic development embedded in them – have become lightning-rods charged with controversy relating to a broad spectrum of issues concerning human/environment relations (White 1995). The fierce battles that have raged over the Sardar Sarovar Project in India and the Three Gorges Dam Project in China are dramatic cases in point (Morse and Berger 1992). But nowhere have these controversies been more sharply etched and widely publicized than in the James Bay Project in Northern Quebec. Begun in the early 1970s as a means to propel economic development at the provincial level by providing an abundant supply of energy, the project not only became a symbol of the increasing assertiveness of Quebec but it also spilled over to affect both relations between Quebec and the rest of Canada and transboundary relations with the United States (Bourassa 1985). At the same time, the project became anathema to those concerned about the environmental and social impact of efforts to transform biophysical systems on such a large scale in the interests of promoting human welfare. Regarded by many as "the plot to drown the north woods," James Bay came to exemplify what an array of critics now see as unsustainable development in the world's advanced industrial societies (Richardson 1972). Starting with a protracted battle over the La Grande project during the 1970s, supporters and opponents of James Bay energy development clashed again during the 1980s and 1990s over the proposed Great Whale project, which was intended to add 3,168 kilowatts of power to the 15,732 kilowatts that will be generated by the La Grande project when it reaches its full potential. Put on hold by the provincial government in late 1994, largely for economic reasons, the Great Whale project lies dormant today. But it could re-emerge and re-ignite the emotional feelings associated with large hydroelectric projects at any time in the future. The furore touched off in late 1997 by reports that Hydro-Québec and the Quebec government are considering plans to supplement water levels in the La Grande complex by diverting water from the Rupert River by way of the Eastmain River clearly demonstrates the continuing potency of issues relating to James Bay energy production (see chapter 2).

This book – a work of analysis rather than advocacy – examines the anatomy of the controversy that has swirled around the James Bay project – the La Grande and Great Whale projects combined – from the 1970s to the 1990s and seeks, in the process, to determine whether there are lessons to be drawn from such an analysis that are applicable to other cases as well as to James Bay itself. The contributors are interested, at one and the same time, in finding ways to integrate the knowledge of natural scientists and social scientists to deepen our understanding of human/environment relations and to link science and policy to encourage a productive dialogue between practitioners and scholars in this increasingly important area of enquiry. The book can be read with profit by those whose interests are limited to the case of James Bay. But to the extent that it succeeds in its larger mission, it will be of interest to those concerned with the rapidly growing problems of environmental governance more generally.

James Bay and the Future of Canada

The controversy over James Bay has unfolded during a particularly tumultuous period in Canadian history. It is no exaggeration to say that Canada has experienced greater challenges to its integrity as a unified society during the last ten to twenty years than at any other time since the establishment of Canada as a distinct political entity through the passage of the British North American Act of 1867. In part, this is a matter of coming to terms with an agenda of unresolved issues relating to the status of indigenous peoples and their claims to the land and resources located within the boundaries of present-day Canada. Once dismissed as quixotic by a dominant society secure in its sense of moral as well as political and economic superiority, these claims have surfaced again with a vengeance as self-doubt has emerged within the dominant society and the assertiveness of Canada's indigenous peoples has grown. Today, the "existing rights" of aboriginal peoples are guaranteed under the terms of the Canadian constitution adopted in 1982. Despite the obvious potential for backsliding in the implementation of such vague guarantees, the federal government has responded to continuing pressure and adopted remarkably accommodating policies regarding aboriginal claims to land, political autonomy, and human rights. Most recently, this has resulted in a landmark decision to divide the Northwest Territories through the creation of a new political unit in the eastern Arctic to be known as Nunavut and dominated by a native majority (Canadian Arctic Resources Committee 1993), as well as in federal government efforts to respond affirmatively to the 1996 report of the Royal Commission on Aboriginal Peoples (Royal Commission on Aboriginal Peoples 1996).

Even more striking is Quebec's struggle for recognition as an autonomous or separate society, either within the framework of the Canadian confederation or, failing that, as an independent political entity. The drive to adopt the 1982 constitution and the associated Charter of Rights failed to bind Quebec more securely into a reinvented Canadian political system. In fact, Quebec was the only province that refused to sign the agreement with the federal government regarding the adoption of the new constitution (MacFarlane 1997). Subsequent efforts to come to terms with the business left unfinished by the patriation of the constitution – the plans known as the Meech Lake Accord of 1989 and the Charlottetown Accord of 1993 – failed to heal the rift. On the contrary, the intense conflicts surrounding these accords coupled with their highly publicized defeat undoubtedly fanned the flames of separatism in Quebec. The latest episode in this ongoing drama, the closely fought 1995 referendum in

Quebec, not only failed to resolve the issues at stake; it left all sides with a sense that new and possibly sharper battles lie ahead. The ultimate resolution of this threat to the integrity of Canada is unpredictable today. But it is beyond doubt that the future relationship between Quebec and the rest of the confederation is the overriding political issue facing Canada at this time.

Any attempt to understand the James Bay controversy without reference to these larger issues concerning the future of Canada is bound to fail. Planned initially with little concern for the probable environmental and social impacts of large-scale energy development in a northern landscape, the James Bay project soon became a battleground for those concerned with the rights of the indigenous inhabitants – the Cree and Inuit – of this vast area. This resulted, in the first instance, in a recognition of the validity of the land claims of the aboriginal peoples of northern Quebec and the negotiation of the James Bay and Northern Quebec Agreement (JBNQA) of 1975, the first modern, comprehensive claims settlement between indigenous peoples and federal and provincial authorities in Canada. Regarded today as a somewhat conservative document, the JBNQA established a tripartite land regime under which the Cree and the Inuit receive "a number of rights, benefits and services in exchange for the extinguishment of their Indian title," together with a monetary settlement totaling $275 million (Moss 1985, 686). Equally important, the agreement created a number of administrative entities and commercial organizations that have become driving forces in empowering the Cree and Inuit of northern Quebec by giving them the resources to pursue their interests with vigour in a variety of public forums. Under the circumstances, the battle of the 1980s and 1990s over the Great Whale phase of the overall James Bay project involved a substantially different balance of forces than the battle over the La Grande phase during the 1970s.

At the same time, James Bay looms large in the deepening crisis of the 1980s and 1990s associated with Quebecois separatism. Hydro-Québec, the organization responsible for generating and distributing power from the hydroelectric facilities of the James Bay region, is a public corporation created in the 1960s through the efforts of René Lévesque and owned by the Government of Quebec. So also was the Société d'énergie de la Baie James, which handled the actual construction of the dams, generators, and related infrastructure of the La Grande Project. The success of the James Bay project has become a highly visible indicator of the growing economic strength of the province. In the minds of many, it stands as a symbol of the proposition that Quebec could prosper economically on its own and without the economic benefits flowing from membership in the Canadian confederation. What is more, Hydro-Québec evolved as a predominantly French organization in a province whose business sector had long been an Anglophone stronghold. The corporation has played a prominent role not only as a source of the energy needed to fuel economic growth in the province but also as a training ground for a Francophone élite whose members now occupy many important roles in the province. Small wonder, then, that the welfare of Hydro-Québec has been a high priority for both Liberal and Parti Québécois governments in the province, and that attacks on the James Bay project itself invariably fan the flames of Quebecois separatism in a political environment that is already exceedingly sensitive.

James Bay as a Transboundary Issue

Most studies of international environmental affairs focus either on international commons, like the high seas, geomagnetic spectrum, or stratospheric ozone layer, or on shared natural resources, like rivers that flow through two or more countries, or animals that migrate across national boundaries. Important as these matters undoubtedly are, they should not deflect our attention from another, rapidly growing class of transboundary environmental concerns. These are efforts to exploit natural resources or use ecosystems lying within the jurisdiction of a single state that nonetheless have significant consequences for the interests of those located elsewhere. As the rise of interdependencies tightens the bonds connecting individual members of international society, these spillover effects of activities nominally contained within a single jurisdiction grow; the result is surely the most rapidly expanding component of the overall agenda of international environmental concerns today. Much has been written, in this context, about the consequences of the destruction of moist tropical forests in countries like Brazil for the protection of biological diversity and even for climate change. But nowhere is this increasingly important class of international environmental concerns more clearly posed than in the case of transboundary issues that arise from the continuing evolution of the James Bay project and affect a variety of constituencies in the United States.

At one level, this phenomenon is a matter of economic links. For some time, Quebecois proponents of continuing hydroelectric development in northern Quebec have argued that James Bay power will find a ready market in the American Northeast as well as in Quebec itself. Initially, this argument seemed compelling; Maine, New York, and Vermont, for example, all negotiated long-term contracts with

Hydro-Québec for sizeable deliveries of electricity. At the same time, much of the capital required to finance the James Bay project has come from American investors convinced that the demand for electricity will continue to grow during the foreseeable future, and that hydroelectric power will look more and more attractive as calls for decarbonization intensify. Yet these calculations now seem highly questionable. Driven by a variety of forces, including but not limited to public opposition to the James Bay project, Maine, New York, and Vermont cancelled their contracts with Hydro-Québec. Investors faced with the prospect of a highly publicized controversy, moreover, began to rethink their commitment to Hydro-Québec; some even bowed to the initiatives of vocal opponents of James Bay and divested themselves of their holdings in the corporation to avoid having to take a stand on the divisive issues raised by the project.

At another level, the internationalization of the James Bay project rests on political concerns. A succession of policy makers in Quebec have welcomed the growth of energy-related ties to the American Northeast. Always interested in developments that would increase Quebec's autonomy *vis-à-vis* the rest of Canada, provincial leaders came to believe that transboundary links with the United States would increase their bargaining strength within Canada. Conversely, American critics of James Bay have long suspected the project of serving as a means of maintaining the energy-intensive lifestyle of Americans while giving rise to a kind of eco-imperialism by deflecting the environmental costs of this way of life onto others. To them, James Bay became a paradigm of the unwillingness or inability of the United States to deal with its own environmental problems, even as it called on others to protect biological diversity or reduce carbon emissions. Under the circumstances, James Bay emerged as much more than an "ordinary" megaproject. It acquired a powerful political valence at the international level as a symbol of the Quebecois search for national identity and the exercise of American dominance, two of the most sensitive and divisive issues on the international political agenda today.

Although some of the consequences of the emergence of the James Bay project as a transboundary issue are diffuse and difficult to capture in any objective way, others are direct and unambiguous. Should government agencies, like departments of environmental conservation or public utilities commissions in Maine, New York, or Vermont, be required to include transboundary effects in their efforts to fulfil obligations to conduct environmental and social impact assessments? If so, should actors located in other jurisdictions, like Hydro-Québec, be obligated to cooperate in the conduct of such exercises? Should private investors be expected to devise and adhere to explicit principles – such as those now known as the Valdez Principles – that apply to projects located in other countries? Are there limits on the capacity of Hydro-Québec, a public corporation, and American public utilities commissions, relatively autonomous governmental agencies, to enter into what amount to international agreements? Are there legitimate national interests regarding this growing web of transboundary relations between states and provinces, and if so, when do these national interests come into play? There is nothing unusual about the James Bay project in this regard. Rather, it exemplifies with particular clarity the rapid growth of an array of transboundary environmental relationships likely to escape the attention of those whose vision is restricted to the traditional realm of interstate relations.

Tracking James Bay's Environmental and Social Impacts

Those who took the decision to initiate the La Grande project in the early 1970s were under no obligation to make a concerted effort to assess the environmental, much less the social, impacts likely to flow from this megaproject. By contrast, the controversy over the Great Whale project triggered an impact assessment effort of massive proportions that nonetheless has left many of the central issues in the debate unresolved. Five distinct assessment processes were involved, four required under the terms of the James Bay and Northern Quebec Agreement of 1975 and one mandated under the Environmental Assessment and Review Process established by the federal government of Canada. Joined together for purposes of scoping and data collection, these processes recognized the federal government of Canada, the provincial government of Quebec, Hydro-Québec, the Cree Regional Authority, and the Kativik Regional Authority as legitimate stakeholders whose interests must be taken into account in the effort to understand the impacts of large-scale hydroelectric development in northern Quebec. They also acknowledged the need to consider the full range of social impacts as well as the more familiar array of environmental impacts. But this did not make impact assessment straightforward or uncontroversial in James Bay; far from it. Conflict over various aspects of the impact assessment processes became a flashpoint for the overarching James Bay controversy (Fenge 1992).

Environmental impact assessment and social impact assessment emerged during the 1980s as distinct fields of study with relatively well defined methodologies of their own (Gilpin 1995). The assessment of the environmental and social impacts likely to arise in conjunction with the Great Whale project therefore afforded an attractive opportunity to bring the methods and findings of this new science to bear on a complex policy problem of great current interest. Yet the effort to do so also exposed some of the complications that those who engage in impact assessment regularly confront. It is a major objective of this book not only to apply the techniques of impact assessment to illuminate the issues at stake in the controversy over the James Bay project but also to probe the prospects for improving the quality of impact assessments through an examination of this seminal case. To this end, the authors of the various chapters consider the following concerns.

How can we establish appropriate baselines against which to assess the impact of the James Bay project? Partly, this is a simple, though nonetheless profound and troubling, data problem. There is much we do not know about the physical, biological, and social systems of northern Quebec as they existed prior to the initiation of hydroelectric development in the 1970s. In part, it is a more complex matter of deciding what we mean by the concept of a baseline. In thinking about the impacts of the Great Whale project, for example, should impact assessment look to the conditions of 1973 or to those of the early 1990s as a proper baseline? More generally, does it make sense at all to single out specific points as baselines in systems that are highly dynamic and subject to continuous change? To the extent that such change is largely cyclical in nature, the choice of a particular point as a baseline may actually mask the existence of a recurrent pattern of change and result in misleading conclusions regarding the impact of resource development. Where change is unidirectional, on the other hand, there is a sense in which choosing any particular point as a baseline is arbitrary.

Can we devise convincing methods to demonstrate causal connections between megaprojects like James Bay and their apparent environmental and social impacts? The fact that environmental and social changes follow large-scale resource development is not sufficient to prove a causal connection; the danger of spurious correlations is much too great for that. How can we be sure, then, that hydroelectric development is actually the cause of the environmental and social changes we are documenting? This central methodological problem of impact assessment is much in evidence in the case of James Bay. We cannot conduct controlled experiments regarding such matters, and the differences among individual cases are so great that the application of generalizations derived from the study of other megaprojects as a means of understanding the probable impacts of a project like Great Whale is fraught with difficulties. Nonetheless, we are not without analytic tools that can help us to deal with such matters. It certainly makes sense, for example, to examine the mercury problem resulting from the impoundment of water in connection with the La Grande project in thinking about the environmental impacts that are likely to flow from the Great Whale project (see chapter 4). Similar observations are in order regarding efforts to foresee social impacts on the mixed Cree/Inuit community of Great Whale. More generally, we can make use of counterfactuals in identifying environmental and social impacts and engage in concerted efforts to explain these impacts in terms of rival hypotheses before treating them as accepted consequences of a development like the James Bay project (Fearon 1991; Biersteker 1993).

What is the proper way to deal with interaction effects, cumulative impacts, and non-linear processes in thinking about the impact of the James Bay project? The ecosystem models implicit in much environmental impact assessment now seem inappropriately simplistic; social impact assessment has been hampered by the absence of any workable models at all. Today, it is clear that we cannot come to terms with the behaviour of fish or animal stocks without understanding their place in the ecosystems to which they belong. These systems, in turn, may well be subject to non-linear change, so that seemingly modest perturbations can trigger profound changes leading to the restructuring of whole ecosystems (National Research Council 1996). To this we must add that anthropogenic impacts have become major determinants of change in many physical and biological systems, though the nature of the interactions between social systems on the one hand and biophysical systems on the other are poorly understood in almost every case (Goudie 1994; Vitousek et al. 1997). For their part, social systems are affected simultaneously by a wide range of forces, so that it is not easy to determine what proportion of the resultant social change can be attributed with confidence to the impact of a particular megaproject (see chapter 5). All this has bred a growing sense of caution among those seeking to project the environmental and social impacts of large-scale resource development. Whereas interaction effects and non-linearities may often serve to magnify these impacts, we are more and more aware of the limits of our ability to pin down, much less to predict, their probable course. The case of James Bay constitutes an excellent laboratory in which to explore these issues.

How should we think about the indirect impacts or collateral effects of large-scale resource developments like the James Bay project? As many observers have noted, the development of infrastructure, which is incidental to megaprojects, can trigger processes whose environmental and social impacts ultimately exceed those of the project itself. A classic case in point – much in evidence in the James Bay case – centres on the construction of service roads or haul roads, which authorities find themselves unable to close to other users once they are in place (see chapter 3). The result is an opening up of remote areas both to tourists and to those interested in long-term settlement. Some of the consequences may prove beneficial to the permanent residents of a region like James Bay. Tourism, for instance, may give rise to new industries and employment opportunities; it may even have a significant impact on the production and sale of regional arts and crafts (see chapter 7). In the final analysis, however, the long-term environmental and social impacts of increased access and the influx of population are likely to overshadow any impact attributable to the operation of the project itself. There is a legitimate question regarding how far we should go in seeking to trace second-, third-, and fourth-order impacts of a development like the James Bay project. Causal chains tend to become thin as we move from link to link. Even so, the history of other large-scale freshwater projects makes it clear that it would be a serious mistake to confine our attention to the impact of dams, generators, and the impoundment of water in contrast to the impact of collateral developments like access roads.

What is the proper role of impact assessment in the policy process? Strictly speaking, impact assessments are intended to provide raw materials that will help policy makers to arrive at enlightened conclusions rather than to dictate the choices they should make. But as the James Bay case demonstrates in vivid fashion, the process of impact assessment often becomes highly politicized in its own right. Opponents of megaprojects endeavour to use procedural requirements to drag out the assessment process in the hope that larger economic and political changes will make the projects irrelevant. Proponents, in contrast, regularly seek shortcuts that will speed up the process and that can be promoted in the name of efficiency. What is more, the line between science and politics often blurs in these processes. Much to the dismay of many scientists, both the production and the use of knowledge become politicized in connection with efforts to understand the impacts of large-scale energy projects. Nor are scientists above seizing on requirements relating to impact assessment to generate resources to conduct their own research. It follows that any hope of separating analysis from policy making and moving from one to the other in a rational sequence must be set aside in situations of this sort.

The discussion to this point presupposes that Western scientific reasoning constitutes the appropriate mode of analysis to be used in probing the environmental and social impacts of the James Bay project. Without questioning the value of this mode, however, it is worth asking about other forms of knowledge and, more specifically, about what is often called traditional (or indigenous) ecological knowledge (TEK) as sources of insight into the complex linkages involved in environmental and social impacts (Freeman and Carbyn 1988; Freeman 1989). Even within a Western scientific frame of reference, TEK has much to offer in the conduct of diachronic or longitudinal studies of systems that fluctuate markedly or change dramatically through time. But beyond this, traditional or indigenous knowledge rests on non-Western perspectives on human/environment relations that suggest a distinctive set of questions about environmental and social impact and that give rise to a different set of procedures for advancing and testing knowledge claims. It is not easy to compare and contrast the propositions about impacts flowing from Western scientific and traditional, or indigenous, modes of reasoning. In some cases, there is no meaningful way to translate questions, much less answers, from one form of discourse to the other. But it would be a major blunder to dismiss traditional or indigenous insights regarding environmental and social impacts on the grounds that they are not rooted in Western science. Here, too, the James Bay project is an unusually rich case (see chapter 6). Because the inhabitants of northern Quebec are predominantly Cree and Inuit who have been keen observers of the region's physical and biological systems for many generations, their testimony should carry a good deal of weight in any satisfactory assessment of the probable impacts of the Great Whale project (Sallenave 1994).

Reconciling Rights and Interests in James Bay

Whatever we conclude about the probable impacts of projects like those involving the development of hydroelectric power in northern Quebec, it is often necessary, at the end of the day, to make a definite decision about whether to go forward with a specific project or phases of a larger project. Once the province of a small number of policy makers, administrators, and organized interest groups, decisions of this sort have become far more complex over the last several decades. Partly, this is a result of the mobilization and empowerment of new players. Today, the rights of aboriginal peoples are recognized under the terms of section 35 of Canada's 1982 constitution; the interests of environmental advocacy groups are acknowledged in a number of requirements like those covering environmental assessment and review procedures, and the concerns of outsiders (for example, nongovernmental organizations with transboundary concerns) are inescapable whether or not they are publicly acknowledged. In part, the growing complexity of such decisions stems from the propensity of stakeholders to frame their claims in terms of rights in contrast to interests. Whereas conflicts of interest – which lend themselves to util-itarian calculations such as benefit/cost analysis – typically give rise to a bargaining process featuring mutual concessions, conflicts of rights are often treated as matters of principle that brook no compromise. The assertion of rights is understandably attractive to parties that are weak in terms of conventional economic and political resources. But conflicts in which each of the stakeholders bases its claims on the assertion of rights are notoriously difficult to resolve in a democratic fashion.

It is hardly surprising that traditional procedures for arriving at col-lective choices – negotiation, legislation, litigation – often prove inad-equate to the resolution of conflicts of the sort exemplified by the con-troversy over the Great Whale project (Osherenko and Young 1989). In dealing with problems of this kind, Canadian policy makers have long tended to eschew litigation and to rely more heavily on negotia-tion than their American counterparts. In some ways, this is an attrac-tive feature of Canadian political culture. But negotiation has major drawbacks as a method of coming to terms with such issues. It is far from clear who the parties to such a process should be and whether they can all be accommodated in a single negotiating process. Are the principals in a case like Great Whale merely the federal government of Canada and the government of Quebec? Is Hydro-Québec a stake-holder in its own right or just an arm of the provincial government? What is the place of the Cree and the Inuit in such negotiations? How are the concerns of various environmental groups to be factored into the process? Do American investors and consumers have a legitimate right to participate? Mixed-actor negotiations of the sort required to include all major stakeholders in a case like this are relatively unfa-miliar, especially to agents of governments used to exercising the authority to make decisions and the power to implement and, if nec-essary, to enforce them. Moreover, it is not easy to identify a contract zone in a situation as complex as this, much less to devise a mechanism through which the positions of the parties can converge on the terms of a specific agreement (Raiffa 1982).

This should not be taken to mean, however, that either legislation or litigation offers a way out of such problems. Quite apart from the fact that these processes are less compatible with Canadian political culture than negotiation, both have drawbacks that are even more serious than those associated with negotiation in dealing with situa-tions like the controversy surrounding the Great Whale project. There is, to begin with, the problem of jurisdiction. Is there any persuasive reason to opt for either the provincial or the federal parliament as the proper forum for arriving at a decision regarding a project like Great

Whale? In either case, how can the concerns of those with little or no representation in these parliamentary settings (for example, the Cree and the Inuit or American consumers) be taken into account? Similar observations are in order about the potential role of provincial courts, federal courts, and even the International Court of Justice. Which stakeholders would have standing in one or another of these judicial forums? Whose interests would be privileged by a decision to proceed in any one of these settings? Even if the problem of jurisdiction could be solved in some satisfactory manner, it is by no means clear that a legislative or a judicial body could resolve controversies like those associated with the Great Whale project or the overarching James Bay project in any decisive fashion. Given the degree to which the various stakeholders typically mobilize to defend their rights and interests, there is little chance that any outcome unacceptable to one or more of the major stakeholders could be implemented successfully, regardless of its formal adoption by the relevant parliament or court.

Where does this leave us with regard to reconciling the rights and interests of those concerned with the Great Whale project? The fact that the battle over this matter developed into a protracted and costly one is no cause for surprise. In situations where many diverse stakeholders are fully mobilized and prepared to go to extremes in support of their positions, blocking coalitions are easy to assemble and policy gridlock is a predictable outcome. Ironically, the government of Quebec postponed the Great Whale project indefinitely – largely for economic reasons – in the fall of 1994, thereby making it possible to avoid any final resolution of the controversy, at least in the short term. But this does not put an end to the long-running saga of James Bay overall. The Great Whale project can and very likely will resurface on the political agenda sometime in the not too distant future. In any event, James Bay exemplifies a class of policy issues now emerging in many parts of the world. Accordingly, a study of the advantages and disadvantages of various approaches to resolving the central issues embedded in the James Bay controversy is important not only in its own right but also as a means of illuminating a range of problems that policy makers everywhere will be compelled to deal with on a regular basis in the years to come.

There are no simple solutions to these problems. Yet several principles that should inform efforts to come to terms with them are already clear. It is important to involve all major stakeholders in any process aimed at solving controversies like the one associated with James Bay; interested parties are likely to accept outcomes – whatever their content – to the extent that they have a sense of ownership of the product that derives from playing an active role in the process. In cases like this, where the parties are bound to engage in a continuing relationship lasting over a long period, it is critical as well to arrive at solutions that seem fair or equitable to all the major stakeholders. Those who feel exploited in such matters are seldom content to play their roles in the ongoing relationship without resentment; they are apt to seize each and every opportunity to reopen the underlying issues at stake. Beyond this, there is much to be said for making a concerted effort to bring a number of techniques of problem solving, or what some commentators call alternative dispute resolution, to bear on matters of this kind. These techniques, which often involve roles for one or more third parties, range from conventional notions of conciliation and face saving to more complex and innovative efforts to open opportunities for mutually rewarding tradeoffs by joining issues together creatively, building mutual understanding among individuals or subgroups operating within the major stakeholders, redefining the nature of the underlying problem, and fostering social learning on the part of all the participants (Susskind and Cruikshank 1987). The resultant processes are apt to be messy; they will not appeal to those whose thinking is tied to simple models of negotiation, legislation, and litigation as methods of coming to terms with social conflict. Yet a capacity to work comfortably with such messy processes will be critical, in the future, to escaping protracted and costly gridlock regarding such issues as James Bay, which centre on complex human/environment relations.

James Bay and the World Energy Equation

Finally, the James Bay project brings sharply into focus the central debate of our time concerning the political economy of energy as well as the influence of competing discourses that structure thinking about this subject (Dryzek 1997). On one side stand those who argue that the demand for energy is bound to rise as human populations increase and more and more countries experience industrialization; that the case against increased reliance on fossil fuels or nuclear energy is overwhelming on environmental grounds; and that hydroelectric power represents a comparatively clean and otherwise benign alternative source of power. Opposed to them are those who maintain that the environmental and social impacts of hydroelectric development are unacceptable in their own right; that there is enormous potential for controlling or even reducing demand through various forms of energy conservation and cogeneration; and that dematerialization offers a viable means for cutting back on the consumption of energy over time. What are we to make of this dichotomy, and where does the battle over

James Bay hydroelectric power fit into this picture? Those who adopt the first view are apt to endorse the James Bay project on the grounds that the use of hydroelectric power on a large scale constitutes the best available option, even though they do not deny that this strategy has environmental costs. Their opponents can be expected to come out against the James Bay project and others like it on the grounds that the option of relying on hydroelectric power merely delays the day of reckoning regarding energy, while at the same time causing immense environmental harm and violating the human rights of indigenous peoples or other local inhabitants.

We do not pretend to be able to reconcile these divergent perspectives in this book. Yet we believe it is possible to contribute in a modest way to the effort to come to terms with them. There is much to be said, for example, for making a careful attempt to sort out and evaluate the inflammatory claims and counterclaims regarding probable environmental and social impacts that have been voiced at various times during the long-running controversy over James Bay. Similarly, we can strive to devise processes for making decisions that give all legitimate stakeholders a meaningful voice and offer some hope of producing results that are widely accepted as fair or equitable. The world energy equation cannot be solved in any synoptic fashion; the overall solution will emerge incrementally from a continuous stream of choices framed in terms of specific issues like James Bay. Our best hope lies in finding ways to make these specific choices in an informed and sensitive fashion.

REFERENCES

Biersteker, T.R. 1993. "Constructing Historical Counterfactuals to Assess the Consequences of International Regimes." In V. Rittberger. *Regime Theory and International Relations.* Oxford: Clarendon Press: 315–38.

Bourassa, R. 1985. *Power from the North.* Scarborough, ON: Prentice-Hall.

Canadian Arctic Resources Committee. 1993. "Creating Nunavut and Breaking the Mould of the Past." *Northern Perspectives* 21, no. 3: 1–18.

Dryzek, J.S. 1997. *The Politics of the Earth: Environmental Discourses.* Oxford: Oxford University Press.

Fearon, J. 1991. "Counterfactuals and Hypothesis Testing in Political Science." *World Politics* 43: 169–95.

Fenge, T. 1992. "The Great Whale Review and the Federal Government's Disappearing Act." *Northern Perspectives* 20, no. 2: 2–5.

Freeman, M.M.R. 1989. "Graphs and Gaffs: A Cautionary Tale in the Common-Property Resources Debate." In F. Berkes. *Common Property*

Resources: Ecology and Community-Based Sustainable Development. London: Belhaven Press: 92–109.

Freeman, M.M.R., and L.N. Carbyn, eds. 1988. *Traditional Knowledge and Renewable Resource Management in Northern Regions.* Edmonton: Boreal Institute of Northern Studies.

Gilpin, A. 1995. *Environmental Impact Assessment: Cutting Edge for the Twenty-First Century.* Cambridge: Cambridge University Press.

Goudie, A. 1994. *The Human Impact on the Natural Environment.* Cambridge, MA: MIT Press.

MacFarlane, S.N. 1997. "Sovereignty and Stability: The Domestic and Regional Security Implications of Quebec Separation." An occasional paper from the Dickey Center, Dartmouth College, Hanover, NH.

Morse, B., and T. Berger. 1992. *Sardar Sarovar: Report of the Independent Review.* Ottawa: Resources Futures International.

Moss, W. 1985. "The Implementation of the James Bay and Northern Quebec Agreement." In B.W. Morse. *Aboriginal People and the Law: Indian, Métis and Inuit Rights in Canada,* 684–94. Ottawa: Carleton University Press.

National Research Council. 1996. *The Bering Sea Ecosystem.* Washington, DC: National Academy Press.

Osherenko, G., and O.R. Young. 1989. *The Age of the Arctic: Hot Conflicts and Cold Realities.* Cambridge: Cambridge University Press.

Raiffa, H. 1982. *The Art and Science of Negotiation.* Cambridge, MA: Harvard University Press.

Richardson, B. 1972. *James Bay: The Plot to Drown the North Woods.* San Francisco: Sierra Club.

Royal Commission on Aboriginal Peoples. 1996. *Report of the Royal Commission on Aboriginal Peoples.* Ottawa: Canada Communication Group-Publishing.

Sallenave, J. 1994. "Giving Traditional Ecological Knowledge Its Rightful Place in Environmental Impact Assessment." *Northern Perspectives* 22, no. 1: 16–19.

Susskind, L., and J. Cruikshank. 1987. *Breaking the Impasse: Consensual Approaches to Resolving Public Disputes.* New York: Basic Books.

Vitousek, P.M., H.A. Mooney, J. Lubchenko, and J.M. Melillo. 1997. "Human domination of earth's ecosystems." *Science* 277: 494–9.

White, R. 1995. *The Organic Machine: The Remaking of the Columbia River.* New York: Hill and Wang.

2 Hydroelectric Power Development at James Bay: Establishing a Frame of Reference

STANLEY WARNER AND
RAYMOND COPPINGER

The James Bay project in northern Quebec ranks among the world's largest hydroelectric developments, comparable in scale with such proposed developments as the Narmada dam in India, the Three Gorges dam on the Yangtze in China, the Challawa Gorge dam in Nigeria, and the Xingu project in Brazil. Endeavours of such scale have become shrouded in a common curtain of controversy. Their merit as instruments of economic development is challenged; their intrusion on local or indigenous people – people who are not the principal beneficiaries of the proposed development – is protested; and their substantial restructuring of the landscape produces a wide-ranging set of environmental concerns. There are those who argue that all such megaprojects are inherently flawed and that only small-scale, run-of-the-river hydroelectricity should be included on the checklist of safe, renewable energy forms (Marcus and Goodman 1990). Yet in the arenas of economic development, social dislocation, and environmental impact, each project differs markedly. Simple categorical judgments tend to substitute moral assertions for critical inquiry.

This chapter introduces the James Bay project as a whole and identifies a range of contexts – environmental, social, and political – that help define the contours of analysis and situate the specific issues in the chapters that follow.

One-third of the world's fresh water is found within Canada and half of that (sixteen percent of the world total) is in Quebec. Canada ranks first among nations in the production of hydroelectric power (the

United States and the former Soviet Union follow), and of Canada's 1990 total generating capacity of 305 billion kilowatt hours, fifty-seven percent (175 billion kilowatt hours) is located in Quebec (Bacon 1992). On the eastern shore of James Bay nine major rivers and their tributaries form a combined drainage area nearly the size of France. Hardly household words, two of these great rivers, the La Grande and La Grande Baleine (or Great Whale), have become increasingly known to the public as protest against further dams coalesces.

As elsewhere in this book, the several phases of hydroelectric development in northern Quebec will be referred to as the James Bay project. While this conforms with much of the international reference, both in the popular press and the scientific literature, it is not altogether accurate. Not included under this designation are numerous hydroelectric projects on rivers in Ontario that also flow into James Bay. As well, the Grande Baleine River empties not into James Bay but into Hudson Bay farther north. Pierre Turgeon (1992) suggests that the region of development be called "Radissonia," after Pierre-Esprit Radisson, one of the cofounders of the Hudson's Bay Company in 1670 and the namesake of the Hydro-Québec operating community of Radisson. For the purposes of this study the James Bay Region is that territory of northern Quebec specified in the 1975 agreement with the Cree people.

When the James Bay project was first announced in 1971 northern Quebec was regarded by many primarily as a "bush" wilderness, holding the potential for yet another Canadian chapter of economic growth through natural resource exploitation. To the traditional activities of mining and timber cutting would be added a bold plan for state-coordinated growth centred on energy intensive industries and offering, in addition, cheap baseload electric power for the general population.

No prior environmental impact study had been undertaken; indeed, there was no general presumption in Canada at the time that such studies were required. The formal notion of a social impact study was even farther on the horizon. The 5,000 to 6,000 Cree in the region were scattered in small bands and regarded as incidental. After concerns from several quarters were raised, a joint federal-provincial working group was appointed, not to weigh the acceptability of the project but to recommend which phase should go forward first. The original intent had been to interconnect three rivers in the southern third of the region – the Nottaway, Rupert, and Broadback – into a unified generating system. (The maps on pages xvii–xix give a sense of the configuration of these rivers and their relationship to the Cree communities.) The Groupe de Travail Fédéral-Provincial, as it was called, recommended against this strategy on the grounds that the area had a higher biological productivity, with greater environmental consequences, and more significant problems in the resettlement of the Cree (Rosenberg et al. 1987). As well, rock and soil conditions in this area were found to present a more costly approach than was originally expected. The group recommended, instead, that the La Grande River option be pursued first – an option that at the time (the early 1970s) included a plan to divert water not only from the Eastmain, Petite Opinaca, and Caniapiscau Rivers but also from the Grande Baleine, 210 kilometres to the north. The latter diversion was eventually dropped from the plan to become a separate proposed phase of development – one that has most recently been at the centre of debate.

It is important to bear in mind that the La Grande phase of the James Bay project is now virtually complete. It represents sixty-seven percent of the total area to be flooded if all phases of development are completed and fifty-seven percent of the total proposed megawatts of power. The first of five large reservoirs in the La Grande system began to fill in 1978, the second in 1981. In 1980 all of the Petite Opinaca and ninety percent of the Eastmain River were permanently diverted to the La Grande. The Cree village of Eastmain is situated where the remaining ten percent of the Eastmain River joins James Bay. Completing the La Grande complex is the Caniapiscau Reservoir, 580 kilometres inland. This sprawling reservoir captures forty percent of the headstream of the Caniapiscau River (which flows north towards Ungava Bay) and diverts this water west, into the La Grande system. The combined effect of all the diversions is approximately to double the flow of the La Grande to a long-term average rate of 3,400 cubic metres per second.

Political Overview

The continuing struggle over expansion of hydroelectric development in northern Quebec is enmeshed in several layers of political context. Four, in particular, actively shape the debate. First, it is historically the case that the individual Canadian provinces exercise substantial autonomy in many arenas, including natural resource management and economic development. The scope for federal initiatives or intervention in such areas is decidedly more limited than in the United States, for example, or most of Europe.

Second, the debate over hydroelectric development is caught within Canada's ongoing constitutional crisis. Given the recurring uncertainty surrounding Quebec's separation and independence, the Canadian federal government has been reluctant seriously to question – much less challenge – this particular project.

Third, the relationships between Canada's federal and provincial governments and its indigenous peoples continue to evolve. Within a comparative international context, Canada wins praise for its leadership and initiatives in "helping" native peoples. Critics argue that these efforts are too often historically grounded in a protective welfare model, reflecting a legacy that insufficiently and reluctantly encourages autonomy and self-determination. At the root of that reluctance is the struggle over land rights, where control of these very large territories is seen both as a form of economic capital and an essential part of national spacial hegemony. Present efforts to resolve land claims walk a delicate line, seeking a framework of "nation" within nation that increases autonomy without granting full sovereignty. Nor is it clear that full sovereignty is the outcome desired by indigenous peoples themselves, given present patterns of encirclement, the extent of their economic interdependence, *vis-à-vis* hydroelectricity, forestry, and mining developments in the region and the size of governmental transfer payments.

The three elements just noted combine to shape the fourth: the political process of environmental and social impact assessment in Canada. While the provinces largely control natural resource use and development, the federal government, through the Department of the Environment or the Canadian Environmental Assessment Agency, may assert shared jurisdiction (or an interest) if, for example, the proposed project has an impact on navigable waterways, migratory species, or native peoples. However, the typical approach has been one of delegated responsibility. Broad guidelines are established by the governments and parties involved. The actual collection of primary impact data is principally carried out by the builder of the project, in this case Hydro-Québec and its companion development arm, Société de développement de la Baie James. The usual question of the fox guarding the hen-house arises, although Hydro-Québec has generally received high marks for its field and laboratory work (Rosenberg et al. 1987). Published research on comparable development projects is drawn upon where appropriate. Without undertaking an independent analysis, the federal government may simply certify that the impact study conforms to required standards. Public hearings are held, with opportunities for disagreements or modifications.

Within Canada this entire environmental review process is usually governed by a concept of "mitigation," rather than yes/no decisions on whether to proceed. A notable exception occurred in British Columbia, where the proposed mining of $8.5 billion in copper and gold deposits was blocked on the grounds that it was incompatible with the ecological integrity of the region; instead, a 958-thousand-hectare preserve called the, Tatshenshini-Alsek Wilderness Park was created (*Globe and Mail*, 23 June 1993). Very large projects are understood to have thousands of separate impacts at different specific locations. Building long stretches of road or transmission lines, quarrying for hundreds of retaining walls and dikes, redirecting sections of river during construction, siting intakes, penstocks, tailraces, and spillways – each of these activities generates a long list of impacts that can be documented and ranked from major to minor. Where a fish spawning bed is at risk from a roadway, or a quarry threatens erosion, mitigating steps are sought to avoid or lessen the impact. Mitigation becomes the enabling framework for shepherding large projects forward.

But several classes of impact are thorny or intractable under a mitigation regime. If the project endangers a specie's survival, if it alters the biological productivity or composition of species too severely, if it destroys unique "special places" that are deemed to have aesthetic or historic value that is difficult to assess, and if, as in more recent debates, it can be shown to erode or undermine the culture or way of life of indigenous people, then the framework of mitigation is challenged. One path is simply to weigh the cost of unmitigated harm against the overall value of the benefits. Canadian law invokes a concept of "balance of convenience" to justify public domain interventions that disadvantage some for the larger benefit of others. In November 1973 the Quebec Court of Appeal cited the "balance of convenience" doctrine in overturning a lower court injunction, won by the Cree one week earlier, that had blocked development of the La Grande project. Unable to prevent the project from going forward, the Cree entered into negotiations that eventually produced the James Bay and Northern Quebec Agreement (JBNQA) of 1975.

On 31 August 1993 Hydro-Québec submitted its environmental impact statement on phase two, the proposed Grande Baleine installation (Hydro-Québec 1993). Within its 5,000 pages hundreds of mitigating steps are outlined. Of the various aquatic and continental life forms, several species are classified as rare but none is found to be endangered. The possible impact on a variant of seal adapted to certain inland freshwater lakes was averted by Hydro-Québec's decision not to divert the Nastapoka River as part of the Grande Baleine project.

The Three Phases of James Bay Development

Now nearly complete, the La Grande complex consists of nine major dams, 206 dikes, 37 generating units, and thousands of kilometres of transmission lines. The total water surface area is over 15,000 square kilometers, of which seventy-two percent represents newly flooded land. The La Grande system increases the water surface area of the entire watershed from about fourteen percent to about twenty percent. In comparative terms it ranks among the largest hydroelectric generating complexes in the world, similar in scale to the Columbia River hydroelectric system in the United States, both in water flow volumes and megawatts of power produced (see pages xvi–xvii). In addition to its size, the La Grande project is exceptional in its reconfiguration of a region's waterways – diverting and redirecting multiple rivers over 176,000 square kilometres of wilderness landscape. "Wilderness" perhaps, to people from the South; but homeland for the Cree.

Farther north, straddling the 55th parallel in a sparser land of taiga, flows the Grande Baleine River – the proposed site for the second stage of the James Bay project. Intense political protest by the Cree supported by a number of environmental groups centred mostly in the United States, together with the coming to power of the Parti Québécois, led to a decision in 1994 to postpone this phase of development indefinitely. In scale, the proposed undertaking is physically much smaller than the La Grande complex. One-fifth as much power (3,168 megawatts) and one-fifth the reservoir surface area (3,576 square kilometres versus 15,613 on the La Grande) would be created. Moreover, fifty percent of the total reservoir surface area would be newly flooded land, compared to seventy-two percent for the La Grande.

The significance of the Grande Baleine proposal lies less in its size than in its moment in history, which is marked by increasing doubts about the wisdom of human endeavours on the scale of the James Bay project in all three of its phases. It is a time when demands for environmental protection more successfully challenge economic development projects, and when the rights of indigenous people are acknowledged as being acutely salient. The Cree have forged a political alliance with environmental groups, including the Audubon Society, Greenpeace, Friends of the Earth, the Natural Resources Defense Council in the US, and local action organizations, to take their opposition into a larger international arena of public debate. The struggle to block construction of the Grande Baleine was grounded less in a specific impact analysis of this particular river system and more in a generalized reassessment of the entire James Bay project. Overall, the political protest movement has had greater visibility and strength in the United States than in Quebec (or the rest of Canada). Opposition groups were formed, bills constraining power imports were introduced in state legislatures, rallies and demonstrations were held, condemnatory full-page advertisements appeared in the *New York Times* and other major papers, and Dartmouth and Tufts universities responded to campus protests by divesting themselves of their Hydro-Québec bonds. The Cree, however, continue to hold $63 million in Hydro-Québec bonds in their capital fund; the money was part of the compensation paid under the 1975 JBNQA agreement (Braile 1994).

Internationally, the James Bay controversy is part of the larger question whose environmental standards shall prevail. In the negotiation of the North American Free Trade Agreement (NAFTA) there was a failed attempt to devise common environmental (and labour) standards; in the end, the parties to the agreement fell back on the notion that each country should obey its own laws. That this is an inadequate solution can be seen, for example, in the discrepancy between acceptable levels of mercury in commercially marketed fish in Canada (0.5 parts per million) and the US (1.0 parts per million), although both fall short of the World Health Organization's recommendation of 0.3 parts per million. To take another instance, the World Bank, sensitive to the criticism that its development projects lack adequate environmental assessment, made loans for India's large Narwada Dam project contingent upon additional environmental and social review. India decided to proceed without World Bank funds. And as noted, state governments are considering legislation that would ban the purchase of electric power from Canadian projects that had not passed the same environmental review procedures applicable in the United States. Since Canada remains the largest trading partner of the US, the issue of binational environmental commonality will remain on the table.

The third phase of development – the Nottoway-Broadback-Rupert (NBR) project – was originally conceived as a stand-alone generating system. Located in the southern third of the James Bay region, the project envisioned a reservoir nearly twice the size of the Grande Baleine system, with eight generating stations as opposed to three. In 1997 it became known that Hydro-Québec had initiated new conversations with the Cree, scrapping the original plan for independent generating facilities and instead proposing that a significant portion of the Rupert River be diverted north, into the La Grande reservoir system. The rerouting would be accomplished through a linkage with the Eastmain – Opinaka diversion, which already forms a conduit through to Lake Sakami, and from there into the La Grande. This southern watershed is a region of boreal forest more biologically diverse and productive than the taiga to the north. The Cree communities of Waskaganish, Nemaska, Waswanipi, Ouje-Bougoumou, and Mistissini would all be affected in varying degrees, in terms of lands and waters traditionally used.

Moreover, the market for hydroelectric power from the James Bay region is subject to a number of contingencies. Domestically, Hydro-

Québec has responded to calls for greater energy conservation by instituting new programs of demand management. Internationally, the 1993 decision by New York state not to renew its power import contract with Hydro-Québec, coupled with the political efforts to block other exports to the northeastern United States, brought a heightened measure of uncertainty. Moreover, Vermont's largest utility, Central Vermont Public Service Corporation, had found by 1994 that because of excess capacity it had to sell back to Canada, at a loss, almost all of the imported power it had agreed to purchase from Hydro–Québec under long-term contract. While power exports to the United States have been projected by Hydro-Québec to be only ten percent of the whole James Bay output, it is a particularly coveted ten percent. First, a kilowatt-hour sent to the United States sells for about twice what it does in Quebec; second, future earnings in US dollars help pay off the significant portion of Hydro-Québec bonds sold in the US, serving as a hedge against fluctuations in currency exchange rates. For the moment Hydro-Québec has announced simply that the target date for the NBR phase has been delayed until the year 2007 or later.

The Cree People of James Bay

Only 10,000 years ago the James Bay region was a mile deep under ice. As the glaciers melted and retreated a massive freshwater lake was formed. It later joined with the northern ocean to become what is now the mildly saline James Bay. The broad plain to the east began rebounding slowly (a process that continues today), with inland elevations rising into the 300 to 600 metre range (Pielou 1991). Thousands of lakes and streams now lace the Quebec territory, some flowing west into James Bay and Hudson Bay, some north into Ungava Bay, and some south into the St Lawrence River, creating watershed networks that recycle one-sixth of the world's fresh water.

Part of the larger Algonquin group of indigenous peoples, the Cree began to occupy this territory 4000 to 6000 years ago. (The Cree nation language group extends to the west of James Bay, but here the reference is to the Eastern Cree of Quebec.) Life in this territory was difficult – the land and waters are frozen for two-thirds of the year – and a complex set of skills were honed to negotiate the particular demands of each season. Agriculture was no part of the structure of subsistence. Hunting and fishing constituted the backbone of the Cree way of life, as small groups of two to three families dispersed far into the bush in winter, then reconvened into larger bands in summer, either on the shores of James Bay or at the edges of major inland lakes, such as Mistassini (Tanner 1979). The Cree are the most northerly people of the region to live primarily from the land. The Inuit, whose southern boundary with the Cree is now marked by the mouth of the Grande Baleine River, are a coastal people primarily oriented to Hudson Bay, Hudson Strait, and Ungava Bay.

It is estimated that in 1857, the earliest year for which a population count has been attempted, there were 1,610 Eastern Cree (Tanner 1979, 20). Over the next seventy years there was little growth. But in the 1930s and 1940s Cree communities began to expand significantly, as infant mortality declined and new life-sustaining measures were incorporated into the culture. By 1960 the eight or nine Cree bands comprised nearly 3,000 people. Twice again in the next thirty-five years population would double, reaching about 12,000 by 1995.

During the 250-year period of commercial fur trapping and exchange with the French and English the social relations of production and community life developed a distinctive formation, somewhat akin to the "putting out" system that arose in pre-industrial England (Tanner 1979). But the market demand for furs has withered – in part because of the inroads of animal rights activists – and income from commercial hunting and trapping has declined to under two percent of income from all sources (Salisbury 1986). A new social structure has now emerged, one that sustains hunting and fishing as an important form of household production but that derives an increasing share of income from wage work and transfer payments (Scott 1984).

A modern technology of float planes, skimobiles, outboard motors, rifles and shotguns, and quick-kill traps have increased the efficiency of the hunt. Contemporary housing, electric heat and refrigeration, cable television, and everything from Pampers to Kraft mayonnaise have been blended with traditional patterns of household life. The community infrastructure of schools, health clinics, policing, and public administration is preponderantly under Cree control, within certain broad constraints of federal and provincial funding and regulation. And from the persistent efforts of French and English missionaries, Christian canon and practice have been interwoven with Cree spiritual beliefs, whether in the ceremonies of marriage and death or the annual fly-outs from the bush to observe Christmas and Easter. While some speak of the clash of cultures between indigenous and encapsulating societies, the Cree (and anthropologists) note that the highly visible technological accoutrements of Western life have been incorporated without substantial tension or injury to the integrity of the Cree social order.

To the north of the traditional Cree lands of the over 7,000 Inuit make their home in the coastal areas of northern Quebec. One-fourth live in five communities that are encompassed by the environmental impact study. The Inuit were also signatories to the 1975 James Bay and Northern Quebec Agreement; for them, however, the most significant impact depends on whether the Grande Baleine phase of development is revived. At the mouth of the Grande Baleine River the Inuit village of Kuujjuarapik (population 450), adjacent to the Cree community of Whapmagoostui (population 500), marks the southern edge of Inuit territory. This village would be most affected by environmental changes, including an increase in estuary brackishness, a change in the composition of fish populations, and a significant reversal of water volume flows between summer and winter. (Hydro-Québec 1993, 74, 100–1). One hundred and sixty kilometres up the Hudson Bay coast, approximately three hundred Inuit live in the new village of Umiujak, constructed between 1984 and 1986. A further 250 kilometres north, Inukjuak, a community of nine hundred, is less directly affected by Hydro-Québec's decision not to divert the Nastapoka River as part of the Grande Baleine design. Offshore in Hudson Bay, 150 kilometres from the mouth of the Grande Baleine, lie the Belcher Islands and the village of Sanikiluaq, a community of 550 that was brought into the environmental assessment process only after a political struggle by the Inuit.

Like the Cree, the Inuit have established a more collective political identity than in earlier times, as the need to negotiate their interests with Euro-Canadians has grown. Currently the Kativik Regional Government and Makivik Corporation take on this function, replacing the earlier Northern Quebec Inuit Association. But unlike the Cree, the Inuit have not sought international alliances with environmental groups, nor mounted a campaign in opposition to the James Bay project. They have agreed to a general statement of intent outlining the environmental impact procedures for the Grande Baleine complex, and on 14 April 1994 they signed the Kuujjuarapik Agreement-in-Principle, permitting hydroelectric development in exchange for $30 million in mitigation measures, $100 million (in 1993 constant dollars) in monetary compensation paid over a fifty-year period, and other provisions relating to job training and employment. That agreement became moot with the decision not to proceed with the Grande Baleine project.

Finally, there is one Naskapi village, Kawawachikamach, situated on the perimeter of the study area nine hundred kilometres inland. In the context of the recent controversy very little has been written about this community of 550 people. Since these people were more directly affected by the Churchill Falls hydroelectric project, constructed in the 1950s in Labrador, Kawawachikamach offers the potential for a retrospective examination of social impact.

The 1975 James Bay Agreement

After the 1973 legal suit failed to block construction of hydroelectric dams in the James Bay region, the Cree entered into negotiations over the terms on which development would proceed. For the Cree these negotiations were less than free standing and offered limited recourse since Hydro-Québec had already begun construction. In 1975 the James Bay and Northern Quebec Agreement was signed by the Grand Council of the Cree (Quebec), the Northern Quebec Inuit Association, the Quebec Government, the federal government of Canada, Hydro-Québec, and two subsidiary Quebec Crown corporations established to develop the region. The agreement reaches far beyond the payment of monetary compensation and stands as an unusual document when viewed against the larger history of Canada's relationship to its indigenous peoples. Five aspects of the settlement are especially notable: a three-tiered land regime, a series of community-based commissions and boards, an income security program, an environmental monitoring framework, and a Cree capital fund for long-term economic development.

In the first of these, the immediate lands upon which native communities were built – representing 1.3 percent of the region – were designated Category One lands, their exclusive use and control to be held by the Cree or Inuit. A wider band of lands, surrounding local villages and accounting for another 14.4 percent of the total, was marked as Category Two. There, native people retain exclusive hunting, trapping, and fishing rights but economic development by non-native groups is permitted, provided any lands taken over for such purposes are replaced, or – at Cree option – monetarily compensated. The remaining 84.3 percent of this vast region is defined as Category Three land. The right to hunt, fish, and trap is shared with the general public and economic development is not constrained. Thus, not only is development permitted on 98.7 percent of the territory but the 1975 agreement also stipulates that the flow of rivers on Category Two and Three lands may be altered even if, within some constraints, this produces downstream effects on Category One lands. The Province of Quebec and Hydro-Québec understand the 1975 agreement to provide explicitly for the remaining phases of hydroelectric diversion, damming, and impoundment and therefore argue that in principle, the Cree do not have the legal basis for opposing development, even though issues may be raised within the agreement about appropriate compliance. (See, in particular, section 8 of the JBNQA.) This comprehensive interpretation is disputed by the Cree.

With regard to the second major aspect of the 1975 agreement, the

JBNQA and its companion legislation address a range of issues to do with community infrastructure and control by creating several locally governed boards and agencies. The Cree Regional Authority and Kativik Regional Government were established to represent Cree and Inuit interests, two regional boards of health were founded, and local school boards were established, all of which expanded local control but maintained a framework of provincial and federal regulation and funding. The infrastructure of community water, sewers, electricity, and roads has been enhanced through both the original agreement and subsequent settlements. Police and public security are administered by each village, and an itinerant judicial tribunal tours the communities to hear cases. However, the larger legal regime remains in the hands of Quebec and the federal government. While the Cree and Inuit have negotiated significant provisions in local administrative control, the complex issues of legal standing and economic development rights are still adjudicated within provincial and federal courts. As discussed more fully below, the Cree have augmented the judicial arena through alliances with environmental and political action groups, media presentations, and, on occasion, appeal to the World Court.

An intriguing innovation within the James Bay agreement is the Cree Income Security Program. Each day spent in the bush (with a minimum of 120 days) at winter hunting camps is compensated according to a formula that individually recognizes husbands, wives, and children. The intent was to strengthen traditional subsistence production from the land, particularly since falling fur prices had reduced trapping income to less than two percent of income from all sources. Since 1977 the annual number of participating individuals has fluctuated between 3,740 and 3,043. But for the last ten years the trend has been gradually downward. In absolute terms, the number of participating adults has never been higher, but since Cree family size has been declining, the number of children in the program has also declined. As the total Cree population continues to grow the proportion wintering in the bush continues to decline. The average payment to a family unit was $11,237 in 1991 (Hydro-Québec 1993, 114). Almost assuredly family winter encampment in the bush would have dwindled without the income supplement. A generational decline in traditional knowledge – so common in other instances – has been averted, and supplementary aid from a program by the Cree Hunter and Trapper Association goes towards teaching young children the skills of the bush. At issue is not merely the methods of the hunt but much of the fabric of Cree life – its seasonal rhythms, the communal sharing of work and meat, the traditional standing of elders, and the spiritual understanding of the relationship of life, human and animal, to the land. That said, however, seventy-two percent of Cree do *not* regularly winter in the bush and derive their livelihood from a combination of wage work, transfer payments, and part-time hunting and fishing. As the Cree population continues to expand, the capacity of the land to bear the pressure of increased hunting will become a species-specific issue in future years. Already, the spring and fall goose hunts, which mobilize entire villages, have a noticeable impact on migrating waterfowl populations. (So, too, do the shootings by recreational hunters from Alabama to Quebec, aligned as a gauntlet along the eastern migratory flyway; Boyd 1977; Reed 1991.)

The 1975 agreement also established one of the first examples of an ongoing environmental monitoring framework. As noted, there was no environmental impact study before construction began in 1971. None was required. Indeed, no federal or provincial agency of environmental protection existed; hence there was no systematic baseline data (apart from occasional studies of specific species or biological processes), with the result that few time-change observations against baseline conditions are possible. However, in a "learning-by-doing" context, programs were put in place to quantitatively sample and track wildlife populations, fluctuations in water flows, aquatic productivity, and other ecological processes. Two parallel monitoring regimes were established, one north of the fifty-fifth parallel (and the Grande Baleine River) for the Inuit and Naskapi, another to the south for the Cree. The collection of research data is undertaken by Hydro-Québec and findings are subject to public review by panels of representatives from native groups and Quebec's Environment department.

One of the most significant impacts of constructing these large reservoirs was the unexpected increase in the levels of methylated mercury to six or seven times the typical background values – a cogent reminder that unknown or unintended consequences are always possible in large-scale environmental intrusions. Mercury in this form is ingested by micro-organisms and becomes more concentrated in the higher-level predator fish that feed upon them. For humans who eat significant quantities of fish (in some Cree communities fish constitutes twenty-five percent of the dietary protein), the possibility of mercury poisoning becomes a serious concern. (For a full discussion of the mercury issue, see the analysis by Bill Roebuck in chapter 4.)

The fifth major element in the 1975 agreement was the creation of a Cree development capital fund, to be invested in the usual range of financial assets, with the intent that earnings would support economic and community development projects for generations to come. The Cree negotiated successfully for such a fund after Hydro-Québec made it clear it would not agree to pay an ongoing royalty on each kilowatt of hydroelectric power. The Cree have used the fund to underwrite the Air Creebec plane service, Cree construction companies (especially for building new housing), and other small business and community projects. Contrary to what was hoped, the fund's expenditures have exceeded earnings, bringing some reduction in the capital balance.

Are These Dams Necessary?

One persistent criticism of the James Bay undertaking is the contention that Quebec is profligate in its use of energy, that Hydro-Québec actively promotes the extensive use of electricity, and that an alternative strategy of increased energy efficiency and conservation would eliminate the need to construct such a massive system (Connor-Lajambe 1990; Waldram 1988). That position is both accurate in some of its premises and faulty in some of its reasoning.

Historically, utilities have forecast electricity demand and built capacity to accommodate it. This supply-side approach was reinforced by regulatory pricing that limited the rate of return on a utility dollar of invested capital but permitted the expansion of the total capital base as a means of raising profits. In the early 1980s Hydro-Québec, along with other Canadian power producers, found they had overbuilt capacity – the consequence of several errors in assessing the factors that shape long-term electricity demand. They responded with a three-pronged strategy (Connor-Lajambe 1990; Goodman 1990). Residential and commercial consumers were encouraged to substitute electricity for fossil fuels, particularly in heating. (The proportion of households in Quebec heated by electricity doubled to about eighty percent.) Intensive industrial users of electricity – notably producers of nonferrous metals such as Reynolds Metal, SKW Canada, and Norsk Hydro – were actively courted with internationally competitive discounted rates. Lastly, new electricity export contracts to New York and New England were negotiated at a time when US utilities were experiencing increased opposition to and rising construction costs for both fossil fuel and nuclear installations.

Compared with other nations, Quebec is currently one of the highest users of electricity, in absolute terms, per capita consumption, and electrical input per dollar of national output. (Quebec would rank fourteenth among nations in GDP if it were to secede.) The call to conserve energy arrived late in Canada, in no small part because electricity rates (and production costs) were about half those in the United States (see Table 1, and Howard 1991). In the US local regulators had

Table 1   Comparative electricity prices, January 1993, US cents/kilowatt hour

	Residential Cents/kwh	Commercial Cents/kwh	Industrial Cents/kwh	Industrial as % of Residential
New York	14.5	15.0	10.2	70.3
Los Angeles	10.4	12.7	7.7	74.3
Boston	11.3	10.5	8.0	70.4
Toronto	8.2	8.3	6.7	82.7
Vancouver	5.4	4.8	3.7	68.5
Montreal	5.4	6.7	4.3	79.7
Montreal as % of Boston	48.0%	63.7%	54.3%	

Source: Natural Resources Canada, Electricity Branch. 1993. *Electric Power in Canada 1992,* 24–6. Ottawa: Government of Canada.

Table 2   Demand-side management savings in electrical generating capacity, 1992

	Generating Capacity 1992 (Megawatts)*	Total Installed Demand-Side Savings in 1992 Capacity (Megawatts)	Cumulative Savings as % of Capacity (B % A)
Quebec	31,104	2,870	9.2%
Ontario	33,969	683	2.0%
British Columbia	12,489	75	0.6%
Canada	108,273	4,275	3.9%
Quebec as % Canada	28.7%	67.1%	

Note: "Demand-Side Savings" represents the reduction in required generating capacity derived from energy conservation measures.
Source: Natural Resources Canada, Electricity Branch. 1993. *Electric Power in Canada 1992,* 123, 138. Ottawa: Government of Canada.

*A megawatt is 1,000,000 watts.

begun changing the incentive structure in the middle to late 1980s, permitting utilities to earn a profit on savings accomplished through conservation measures. Such incentives are only now being developed in Canada. Surprisingly to some, the record indicates Hydro-Québec leads the rest of Canada in energy conservation. In 1992 the cumulative effect of Quebec's energy conservation – referred to as demand-side management – was saving the equivalent of 9.2 percent of total installed capacity, compared to a 3.9 percent cumulative saving for Canada as a whole (see Table 2). Quebec accounts for 28.7 percent of Canada's total installed generating capacity but has achieved 67.1 percent of the demand-side savings in power capacity. Much of this has been accomplished through load shifting in peak use periods. Other improvements in the efficiency of electricity consumption have barely been touched and offer the potential for significant future savings. Hydro-Québec has initiated a range of programs, installing electronic home thermostats, working with individual municipalities, and promoting private cogeneration by industrial users.

Sometimes the argument is made that Quebec should not be promoting electricity for such uses as home heating in the first place because it disregards the second law of thermodynamics. That claim is flawed because it applies only to electricity generated from burning fossil fuels. If fossil fuel is burned to produce electricity and the resulting power is used for heating, the second law of thermodynamics instructs that a greater quantity of heat (measured in British Thermal Units) could have been obtained from heating directly with the fossil fuel rather than transforming it into electricity (with the attendant ride up the entropy scale). And since changes in the price of fossil fuel tend to produce proportionate changes in the cost of fossil-fuel-based electricity, what is true in physics is also true in economics. There is no parallel argument in the case of hydropower, since thermodynamics applies only to processes in which heat is converted to work. In more physical terms, there simply is no more efficient way of using hydropower for home heating than through electricity.

The call to replace dam building with a more aggressive energy conservation strategy (through demand-side management) misses an essential dimension of the debate, namely, that one of Quebec's principal agendas is the use of abundant electric power as an instrument of economic development. This expansive strategy views hydroelectric power as a perpetually renewable natural resource that can be used within Quebec to attract new industrial enterprise or be exported from Quebec to the United States as a commodity in its own right. The goal is not to minimize its use but to optimize its development potential.

The critique of Quebec's use of electric power for economic development requires a different set of questions. First, recognizing that demand-side management should continually seek greater energy efficiency, how should all justifiable ongoing and economic development needs be produced: by fossil fuel, nuclear, small- and large-scale hydro, cogeneration, or the "soft," less well trodden paths of wind, solar, tidal, and thermal energy? Second, what is the efficacy of such a development strategy in terms of the combined environmental impact, including that of downstream industrial users, and in terms of the economic ripple effects on jobs and the economy? Third, to whom do these lands (and waters) belong – the province, to be used under the doctrine of "balance of convenience," or the Cree and Inuit, to be reclaimed under a growing movement towards self-determination?

As a promoter of economic development, Hydro-Québec is clearly not the equivalent of a private utility. As a Crown corporation it acts as an agent of Quebec – as Robert Bourassa intended in his book, *Power from the North* (1985), published before his second coming to office as premier. As an agent of the state Hydro-Québec should evaluate the full social costs of its development strategy, including externalities: those additional costs that are born by society but not by Hydro-Québec as part of its private production costs. One of these external effects arises from the nature of downstream industrial development. Among the industrial consumers most sought after have been the smelters of non-ferrous metals, aluminum and magnesium in particular (Goodman 1990). In assessing the efficacy of hydroelectric power as a tool for economic development, one must consider the airborne emissions and other environmental impacts generated by such users. The ripple-through employment stimulus from smelting, moreover, is generally small compared to other forms of industrial development. To illustrate, at Baie Comeau, Quebec, on the shores of the St Lawrence River, Reynolds operates the second-largest primary aluminium plant in the world. Alumina powder brought by ship from Corpus Cristi (Texas), Chile, and other locations is converted to aluminium with large inputs of electric power from the Manicouagan complex to the north and then shipped as ingot to various users worldwide. Beyond the 2,000 jobs created at the plant itself, the secondary and tertiary employment effects are not impressive. By contrast, forestry, whatever its other limitations or concerns, generates more than twice as many direct and indirect jobs per dollar of sales or capital invested. Greenhouse producers of flowers and vegetables also argue that for an equivalent subsidy in low-cost electric power they would constitute a stronger engine for job creation.

The pricing discount given to aluminium smelters and other preferred industrial users has been criticized as a secretly negotiated and unfair subsidy. The discount has been defended on the grounds that these buyers purchase large quantities of off-peak power, carrying a lower marginal cost of production, and on the claim that the social gains from employment creation are correspondingly greater. A third explanation for differential pricing – or "price discrimination," as economists call it – is simply that all utilities practise it for all classes of users. It reflects the power of a monopoly enterprise to charge varying prices based on differences in willingness to pay, or "elasticities of demand." In the debate on the merits of the electricity it produces, James Bay has too often been treated as a special or isolated case, rather than placed in the general context of utility practice.

Should the United States be importing electric power from Canada?

Opponents of Quebec's development strategy have again invoked the demand-management argument, contending that if the US put more effort into conservation, the additional power would not be needed. That linkage entails the same misunderstanding noted above. After demand management has been fully exploited, how should all remaining baseload energy requirements be met? Imported power should not be thought of as the additional or last increment in the equation. During the next twenty years New England will experience significant reductions in its base generating capacity. The Massachusetts Yankee Rowe nuclear plant, with 173 megawatts of power, was recently closed. Other nuclear facilities are scheduled for decommissioning in the next two decades: Connecticut Yankee (560 megawatts), Maine Yankee (870 megawatts), Pilgrim (Boston Edison, 670 megawatts), Millstone One (New England Utilities, 650 megawatts), and Vermont Yankee (520 megawatts). In addition, New England utilities are faced with an aging stock of coal and oil-burning plants that will need to be replaced or substantially upgraded.

In deciding whether to enter into long-term energy contracts with Canadian producers, the United States or its individual state legislatures may take either a localist or a "world citizen" perspective in weighing the full social costs of getting power from alternative sources. The choices require making comparative judgments on such matters as the effects of greenhouse gases associated with fossil fuels and the social impact on native peoples of large-scale hydroelectric development. Finding an acceptable methodology for making such choices – one that brings an interior logic removed from pure "power politics" – is at the cutting edge of contemporary public-choice theory.

It now appears that the job of choosing will be shifted in large part onto electricity consumers. The movement in both Canada and the United States to deregulate the power industry is founded on the premise that ownership of the transmission or distribution network can be separated from the question who produces the electric power, and from what source it comes. Consumers will have a choice among several sources of power that are marketed at varying prices and entail different environmental and social impacts. Such a framework may work to the advantage of Hydro-Québec in its desire to make cross-border sales to the United States. Even if a majority of individual consumers were to choose against Hydro-Québec, the corporation could still meet its target of exporting at least ten percent of production as long as enough industrial, commercial, and other buyers selected simply on the basis of price. Hydro-Québec and other power producers in Canada have restructured their corporations in preparation for this new marketing environment. Will the legislatures of New York,

Vermont, Massachusetts, and other states continue to consider banning cross-border sales by Hydro-Québec, or will they simply leave that decision to individual purchasers? If the states do seek to prevent the importation of Hydro-Québec power, would that constitute a violation of the North American Free Trade Agreement? The weak environmental accord that accompanies NAFTA says only that each nation shall define and enforce its own environmental standards. Some consumers would argue in favour of Hydro-Québec power on the grounds that it is better for the environment than nuclear or fossil fuels, and because they have not been dissuaded by information about the possible impacts on the Cree.

Considered in its totality, the James Bay hydroelectric project submits to no easy assessment of its social implications – ethical or scientific. The enormous reconfiguration of water and land has been underway for over two decades. While this presents the opportunity to learn from changes that have already occurred, it also makes it harder to challenge the momentum of the project. Deciding whether further James Bay development is in the public interest depends on the ability to think through the social gains and losses at several levels of analysis. First, construction of the James Bay generating system cannot be dismissed on the grounds that demand-side management is capable of achieving significant additional reductions in total electricity demand. Second, the social gains from reducing the greenhouse gases associated with fossil fuels or avoiding the constellation of dangers associated with nuclear power are to be counted to the credit of this form of renewable energy; but on the debit side are the environmental impacts that occur when low-cost electric power is used as a development strategy to attract such industries as aluminium and non-ferrous metals smelting. Third, the direct environmental impact on the James Bay region itself falls into many categories, requiring both specific natural-science methodologies for determining particular impacts and an overarching economic and political framework for weighing the combined outcomes. Finally, the Cree people who occupy these lands currently protest further construction of the project on social and ecological grounds, and as part of their quest for greater autonomy and self-determination.

REFERENCES

Bolduc, André. 1992. "A Societal Choice: An Interview with Lise Bacon," Quebec Deputy Premier and Minister of Energy and Resources. *Forces* 97: 4–13.

Bourassa, Robert. 1985. *Power from the North*. Scarborough, ON: Prentice-Hall.

Boyd, H. 1977. "Waterfowl Hunting by Native Peoples in Canada: The Case of James Bay and Northern Quebec." Paper read at the Proceedings of the Thirteenth International Congress of Game Biologists, Washington, DC.

Braile, R. 1994. "Tufts-Hydro Break Sought by Students." *Boston Globe*, 2 February 1994.

Connor-Lajambe, H. 1990. "Societal Impacts of Utility Overinvestment: The James Bay Hydroelectric Project." *Utilities Policy* (October): 78–87.

Goodman, I. 1990. "Hydro-Quebec's long-term export policy." In *Canadian Water Watch*. Ottawa: Rawson Academy of Aquatic Science.

Howard, R. 1991. "Canada Worst in Use of Energy." *Globe and Mail*, Toronto, 15 July.

Hydro-Québec. 1993. *Grande-Baleine Complex Feasibility Study – Summary*. Montreal: Hydro-Québec.

Marcus, W., and I. Goodman. 1990. "Economic and Technical Aspects and Electricity Demand in Quebec". Testimony submitted to the National Energy Board, Hearing Orders no. EH-3-89 and AO-1-EH-3-89.

Pielou, E. 1991. *After the Ice Age: The Return of Life to Glaciated North America*. Chicago: University of Chicago Press.

Reed, A. 1991. "Subsistence Harvesting of Waterfowl in Northern Quebec: Goose Hunting and the James Bay Cree." Paper read at Transactions of the Fifth-Sixth North American Wildlife and Natural Resources Conference, Washington DC.

Rosenberg, D., R. Bodaly, R. Hecky, and R. Newbury. 1987. "The Environmental Assessment of Hydroelectric Impoundments and Diversions in Canada." In *Canadian Aquatic Resources*, edited by M. Healey and R. Wallace. Ottawa: Rawson Academy of Aquatic Science and Federal Department of Fisheries and Oceans.

Salisbury, R. 1986. *A Homeland for the Cree: Regional Development in James Bay 1971-1981*. Kingston, ON: McGill-Queen's University Press.

Scott, C. 1984. "Between 'Original Affluence' and Consumer Affluence: Domestic Production and Guaranteed Income for James Bay Cree Hunters." Paper read at Affluence and Cultural Survival (Proceedings of the American Ethnological Society, 1981), Washington, DC.

Tanner, A. 1979. *Bringing Home Animals: Religious Ideology and Mode of Production of the Mistassini Cree Hunters*. St John's, NF: Memorial University of Newfoundland, Institute of Social and Economic Research.

Turgeon, P. 1992. *Radissonia: The James Bay Adventure*. Translated by Dominique Clift. N.p.: Libre Expression.

Waldram, J.B. 1988. *As Long as the Rivers Shall Run; Hydroelectric Development and Native Communities in Western Canada*. Winnipeg: University of Manitoba Press.

PART TWO

Social and Physical Impacts on the Environment

3 James Bay: Environmental Considerations for Building Large Hydroelectric Dams and Reservoirs in Quebec

RAYMOND COPPINGER AND WILL RYAN

The development of Quebec's hydroelectric resources at James Bay has been highly controversial. Some conservation groups argue that the building of large generating systems could damage the environment irrevocably, devastate indigenous fish and wildlife, and destroy the culture of the Cree people who live in the region. Others argue that development could produce such long-term effects as global warming and, ultimately, the destruction of the entire James Bay ecosystem. One concern is that these long-term effects may be cumulative in that they are not predictable in any linear fashion. This "long-term cumulative" hypothesis outlines neither a methodology that would allow investigators to make predictions about the probable impacts, nor a means for determining the results. It only assumes that cumulative effects are bad, effectively ignoring changes that might actually be beneficial.

Proponents of hydroelectric development insist it is the only practical way to meet the present and future power needs of eastern North America. They believe it could replace oil- and coal-fired electric generators, decreasing substantially the emission of hothouse gases and the resulting acid rain. The impact on the ecology of the region, they say, will be largely benign.

Even with so-called objective methods of science, initial hypotheses often drive research in a direction that reflects the biases of the researchers. To date, most environmentalists concerned with James Bay emphasize the long-range, cumulative consequences and assume that environmental changes are always negative, while "pro-Hydro-

Québec" arguments tend to stress the absence of concrete, identifiable damage. This polarized orientation has done little to advance an understanding of the structural impact of hydroelectric facilities in northern Quebec. To counter alarmist forecasts, Hydro-Québec points to the fact that no negative effects from cumulative changes have been observed. Environmentalists, when asked to respond to this, simply say, "Just wait and see." In effect, both arguments are positioned in ways that cannot be lost.

The immediate environmental effect of the La Grande complex has been the flooding of large areas of taiga, or open forest interspersed with lakes, rivers, swamps, peat bogs, and beaver ponds. Critics, however, are less concerned with the substantial increase in the amount of water in the region than with the system of dams, whose annual operation changes the seasonal discharge and flow patterns of fresh water into James Bay. The natural seasonal rhythm without the dams consists of a tremendous discharge in the spring when the ice melts, tapering to a trickle in the fall, followed by freeze-up. Since Quebecers require higher levels of electricity in the winter, the reservoirs are drawn down during the coldest months, while the spring flood is captured and stored. The flow rate of the La Grande has now doubled; the winter flow has gone from 500 to 5,500 cubic metres per second while the spring flood has decreased from 5,000 to 1,500. In other words, the annual pattern is just about reversed.

Forty percent of the water from the Caniapiscau River and one hundred percent of the waters from the Opinaca and the Little Opinaca (tributaries of the Eastmain River) have been diverted, thus reducing the flow of the Eastmain to a mere ten percent of its former self. This spatial shift in freshwater flow, as well as the temporal shift whereby fresh water is discharged during the winter months because of electricity demand, would also change the total productivity and the species composition of the region.

Thus, several hypotheses emerge. First, changes in species composition result from the flooding of terrestrial and riverine areas (favouring lake-adapted species). Second, changes in the distribution and seasonal discharge of fresh water lead to shifts in the salinity patterns of James Bay, including the timing of ice formation and melting patterns. These changes in turn would change the salinity, productivity, and species composition of the rivers and their estuaries leading into the Bay. Fourth, large fluctuations in reservoir water levels would affect the reproductive success of several species. Moreover, because James Bay is considered by many to be important to migratory birds, it is feared that a reduction in the productivity of the region could lead to declines in migratory populations elsewhere. Finally, these changes, taken together, could warm the world climate by adding the greenhouse gas methane to the atmosphere, while the winter freshwater discharge would change the Labrador current enough to cool those continents bordering the Atlantic Ocean.

In this chapter we consider the overall natural history of the region, review the changes created by the La Grande River projects, and speculate on the impact of the postponed dam on the Great Whale River to the north, and the still more tenuous proposal for the Nottaway, Rupert, and Broadback Rivers to the south. We are offering a first step, not the final word. As it is, the proposed and real environmental changes in the region have led to numerous impact statements, covering thousands of pages, with inventories and observations on particular species habitats. These observations beg the right questions. Our intent, however, is not to assess or judge but to develop a framework that might be useful for making judgments on environmental impact.

Our analysis is limited to environmental questions and their ecological consequences. We recognize that there are consequences for the Cree people in the construction of any hydroelectric facilities, and that they may well be the most important issues. We also realize that there are immense consequences for Quebecers who see hydroelectric power as a developmental resource that allows them to pursue their own economic destiny. Indeed, the human issues demand an assessment study of their own. They are addressed at greater length in chapters 5, 6, and 7 of this volume. In an effort not to confound the various problems, we look primarily at the impact of the hydroelectric projects on the flora and fauna of the region.

Methodology

Ecologists have several methods of classifying and measuring the health of an environment. Typically such indicators of an ecosystem are divided into three major categories: productivity (how many cells the region supports), biodiversity (how many different kinds – species and subspecies – of organisms exist in the region), and species endangerment (which particular species or subspecies are at risk, if any). When considering environmental impact, ecologists ask, for example, whether there will be a significant change in the number of organisms, whether the variety or composition of those organisms will be altered, and whether any particular species will be removed from the project zone. If it is determined that the answer to these questions is substantially yes – a judgment contingent on social values placed on environmental goods – then the project could conceivably be cancelled or modified.

This three-part methodological framework contrasts with the inventories and checklists around which environmental impact statements are often organized. For example, if in a particular locale one species is replaced by another – whitefish for trout, or aquatic organisms for terrestrial – then a change in *composition* has occurred but not necessarily a change in the *total amount* of organic material, nor even in the *total diversity*. In other words, there may be just as many different species even if they do not happen to be the same species as before, and the total biomass might be the same even though a change in the species composition has occurred. Thus, one has to ask whether the trout are actually endangered if the project has altered the region in such a way that trout can no longer live *there*.

Biological production should be viewed in the context of the viability of species within the ecosystem as a whole. The essential issue is the long-term integrity and health of the ecological region. A whole range of changes may occur, whether due to natural causes or human activity, that do not necessarily place the system at risk. This is not to say that the impacts must reach catastrophic levels before they become meaningful. Any change, however gradual, that contributes to the long-term degradation of the health of the biosystem warrants attention.

PRODUCTIVITY

The taiga of Quebec is biologically not very productive, compared with other ecosystems. The area averages only seventy frost-free days a year and photosynthesis is minimal in freezing conditions. Since there is little terrestrial productivity, there is also little growth in aquatic environments; fisheries are often dependent on the nutrients that leach from large land masses into lakes, rivers, and oceans.

Almost all life on our planet is sun-driven. Green plants use the sun's energy to combine carbon dioxide and water into sugar and free oxygen. Plants then use the energy stored in the sugar to extract other elements from the environment to construct the tissue and organs that give them form and structure. Plants are consumed by other organisms to create their own tissue. One way of measuring the potential bioproductivity is to measure how much chlorophyll there is in an area. But the total amount of chlorophyll product has to be factored with the quantity of carbon dioxide and water in the system (both of which can be limiting factors), and of course the productivity of chlorophyll depends on the temperature. The ultimate measurement of gross primary productivity is the rate at which kilocalories of sugar are produced by plants in some defined area. The tundra to the north of the taiga produces 200 kilocalories per square metre a year, whereas the more southerly boreal forest produces 3,000, and a tropical rainforest 20,000 (Chiras 1985). In the taiga, one-hundred-year-old trees are only a few metres tall, ten centimetres in diameter, and sparsely distributed. Our eyes tell us that there are very few cells (biomass) in the system.

Not all areas of the earth receive the same amount or quality of sunlight. Regions differ in diurnal and seasonal quotas of light, and atmospheric clouds and dust particles prevent some sunlight from reaching the plants. Clouds, sand, ice, and snow reflect some proportion of sunlight back into space. Thus, by no means are all the photons of sunlight entering the atmosphere available to plants. In fact, most sunlight energy is converted to heat energy, which warms the surface of the planet, melts ice, evaporates water, and, in effect, drives that meteorological activity we call weather. Sunlight is capable of warming the James Bay region to a temperature where photosynthesis could occur for less than fifteen percent of the year.

If the James Bay project did change the climate, the impact would be significant since primary productivity would be altered. This could happen in several ways. Increasing the amount of water, snow, or ice can increase surface reflectivity (albedo) and thus reflect more of the sun's energy back into space, causing the region to cool. Cooling a very hot region where evaporation is high might increase primary production by making water more available to plants; cooling a cold region (e.g., freezing the water) would decrease photosynthetic activity. A change of just a few degrees of climatic temperature could alter the seasonal availability of water to plants enormously.

The James Bay region receives little sunlight, particularly during winter when, because of the earth's axial position, the region is inclined away from the sun. With long nights and a low angle of sunlight, much of the light is either unavailable or reflected away. Frozen water cannot be used for photosynthesis.

If chlorophyll is mechanically eliminated from a region, then the total biomass will decrease. Bulldozers scraping rocks clean or developments that increase soil erosion prevent plants from carrying out their vital activity.

BIODIVERSITY

Ecologists and resource managers also assess the fertility of a region by considering the diversity of life (Dudley 1992). How many distinct species exist regionally? Theoretically, the more species there are, the more sunlight energy can be captured and stored as biomass.

The principle is that each species is adapted to different habitats (niches) and exhibits the most efficient form for capturing energy in that habitat. Thus, some organisms can extract energy from an aquatic environment, while others require a terrestrial environment. Within aquatic environments some animals have a form (morphology) that allows them to eat plants in shallow fast-moving water, while others may lie in wait at the bottom of a big lake for detritus to settle to them.

The extent of species diversity in any region depends on a combination of factors, not all of which are understood. Uniform environments of low productivity such as the taiga in the northern James Bay region have fewer niches and therefore less species diversity. Species diversity generally decreases with increasing latitude (Wilson 1992). There can be more species of insects, for instance, on a single tree in the tropics than in all of Quebec.

Population numbers in the North are often very large and widely dispersed. Species like the caribou are circumpolar, whereas a tropical species may be found only in a single lake. The species of plants and animals presently found in the James Bay taiga have only recently invaded the region, after the last melting of the glaciers; none are endemic, in the sense that they evolved there or are confined to the taiga alone.

The James Bay taiga is not a distinct zone in the same sense that a tropical rainforest is. Tropical rainforests are large and have been around for millions of years. Species of plants and animals have evolved to those specific conditions that have been stable over long periods. The taiga is a suture zone between the tundra ecozone and the boreal forest ecozone. Taiga forests around the globe are small and fragmented, recent in origin and unstable.

ENDANGERED SPECIES

In assessing the impact of James Bay, or the environmental changes it has caused, the ecozone must be inventoried species by species to ensure that no species is endangered. There are three categories of endangerment: eradicating a species from the earth; eradicating a species locally; and reducing the population density or fragmenting a population to the point where – because gene flow is regionally disrupted – populations tend to become inbred, which is another way of decreasing genetic diversity.

The first and third cases apply to endemic species that this region does not have. Our concern here is the second case, eradicating a species locally. This is often referred to as decreasing the margins. As noted above, in the high latitudes where both the distribution and populations of species tend to be extensive, a development project in the James Bay region is not so likely to endanger a single species as it might be in the tropics, where a project could conceivably wipe out a species across its entire range. It is important to underscore that virtually all species using the James Bay taiga actually have the centre of their populations (greatest population density) elsewhere. Taiga Quebec represents the margin of their range. By altering the habitat the dams may reduce the total size of the species' niche, thus reducing the total population number but certainly not threatening the species with extinction. Although the species' margin contracts, the total world population changes little, since densities are low at the margin.

These three criteria of an ecosystem's health are difficult to apply. The taiga is a land of extremes. Animal and plant populations are cyclical, seasonally and temporally. In the winter and summer there are few waterfowl on the taiga. They appear in the millions, however, for brief periods of resting and refuelling on their way north or south. Caribou herds summer on the tundra ecozone, migrating to the taiga for fall and winter pasturage. When the La Grande project started there were fewer than 200,000 caribou in Quebec; twenty-five years later the population is approaching one million. Wildlife managers predict it will return to a lower level. Environmental impact studies would benefit from data about a population before a project gets underway, and from predictions of what it will be in the future. In the taiga, caribou and most vertebrate species populations are different every year, increasing and decreasing over time.

Many populations of animals in the Far North show these cyclical trends. Beavers (*Castor canadensis*) create artificial ponds and in doing so change the dynamics of the ecosystem (Naiman et al. 1986). They flood and kill large areas of forest every year, changing species composition from spruce to aquatic-loving trees and plants. They feed on aspens and willows. The ponds act as settling basins for soil and organic material. Often within these beaver ponds the rotting vegetation increases the acidity, eventually killing fish and other aquatic life. The rotting vegetation also leads to a substantial increase in methane production (Ford and Naiman 1988), one of the so-called greenhouse gases, and methylated mercury enters the food chain. Primary productivity slows, and species diversity decreases as the area succeeds back to spruce and larch forests. In some areas the cyclicity of beavers is correlated with catastrophic forest fires. Forest fires eliminate the spruce, allowing a succession of events to take place. Birch, willow, and aspen are regenerated and form a food base for beavers. Thus, after a forest fire, species diversity increases and organic material, inert in tree trunks, is now released to become living tissue. Primary productivity increases.

Catastrophic events are a normal part of the ecozone. Fire, flooding, and population blooms and crashes seem to be an integral part of natural cycles and, perhaps, the ecological health of the area.

This formulation suggests another way to judge the health of an area. Recently, the World Wide Fund for Nature started a program to introduce wolves, bears, and lynx into European national parks. Survival of these top-level carnivores, it is reasoned, would suggest that these were healthy ecosystems. Indeed, the presence of predators is an indicator that all the components of a food chain are in place. Comparing the food web before and after a development project would give us some idea about how productivity and diversity have changed, and also whether the change has been healthy or not.

The James Bay Region

PHYSICAL CHARACTERISTICS OF THE JAMES BAY REGION
Located in the centre of North America, the James Bay region falls in the "middle" of the Canadian Shield (see Appendix B). The underlying rock in this area might be some of the oldest on the planet, perhaps two and half billion years old. It consists of fragmented and fractured pieces of old continents. This basement rock is geologically quiet because central portions of continents do not receive the buffeting that is characteristic of continental edges.

But the surface has been disturbed by glaciers over the last few million years. Ten thousand years ago, when people were first developing agriculture in Mesopotamia, the James Bay region was buried under a mile-high block of ice. This last glacier started melting at the time, after its massive movements had scarified the land, grinding the surface and redistributing the debris as moraines and eskers. Cape Cod, for example, is the result of one of its debris-dumping forays.

The James Bay region is still very much involved in these glacial processes. In the recent past a massive block of ice (remnants of which can still be seen on Baffin Island) covered the continent as far south as Long Island, New York. So much of the world's water was frozen in the glacier that sea level was a hundred metres lower than it is today; the present location of the port of New York City was a hundred miles from the ocean, up the river at the head of a canyon that would rival the Grand Canyon. Some eight thousand years ago – a few thousand years before the pyramids were built – the region was still under ice. In fact, the low sea level provided the causeway by which, at the time, the ancestors of the Cree and Inuit crossed from Asia to North America.

The ice was so heavy that it depressed the land under it. As it melted, the land rebounded. In some places the glacial rebound amounted to four hundred metres; it still continues to rebound at a rapid rate in Quebec. As the glacier receded northward, the rebounding land to the south created depressions between the newly exposed land and the glacier. Melting glacial waters filled these depressions, creating glacial lakes hundreds of miles long. Most of the tens of thousands of Canadian lakes today are remnants of much larger lakes formed in this manner.

About seven thousand years ago James Bay itself was one of these large freshwater lakes – the 2,900-kilometre-long Lake Ojibway – separated from salt water on the north by the glacier and on the south by the rebounding continent. But the melting glacier split and the gargantuan lake perched a couple of hundred metres above sea level spilled catastrophically into the sea. As the glacier continued to melt and sea levels rose, Lake Ojibway's basin filled with sea water. Today the bay is still bounded on the south by the rebounding continent; but that rebound is now dumping the salt water back into the Arctic Ocean. (For a detailed description, see Pielou 1992.)

The rebounding process continues at a rate of between 0.9 to 1.2 metres per century – truly a glacial pace in human time, but a rapid rate in geologic time. Both Hudson Bay and James Bay are getting smaller and shallower by the day as they dump their waters into the oceans of the North. Nearly a billion cubic metres of water flow out of James Bay each year because of glacial rebound. This large amount is ten percent of what the La Grande complexes discharge into the bay.

In another four thousand years James Bay will be a giant fen, peppered with freshwater lakes, bogs, swamps, and old beaches. It will look very much like the inland Manitoba and Ontario of today, although the soils will have formed in salt water rather than fresh, producing fenlike characteristics.

The soils of the adjacent taiga are lacustrine, meaning that they formed at the bottom of lakes. Silts and clays, trapped in the glacier, were moved by melting water and settled into the lake bottoms. Because of silting and glacial rebound, lakes gradually became "dry" land. Soils accumulated and changed with exposure to the elements and the action of vegetation and other erosive processes. Since the climate is cold and the land only "recently" free of ice (and not so free from permafrost), very little soil has accumulated. Thus biological growth is further restricted.

The land on the east side of James Bay is in an ecozone called the taiga shield. The southern tip of James Bay, where the Rupert, Nottaway, and Broadback Rivers enter, is in a larger ecozone, the Boreal Shield, that covers much of Canada (and northern Eurasia). Many rivers, including the La Grande, Opinaca, and Eastmain, flow through the taiga ecozone. Ecological changes caused by the construction of the hydroelectric facility on these river systems predict what might happen in other rivers in this ecozone if they were dammed or diverted. The Great Whale River forms the approximate boundary between the taiga and the tundra, an ecozone that receives insufficient photosynthetic light energy and water to support trees.

The taiga shield has an average of only seventy frost-free days a year, and these are not necessarily contiguous. Few days each year have temperatures warm enough to support vegetative growth (photosynthesis). An arid ecozone, the taiga receives less than two hundred millimetres (eight inches) of rain per year. There is neither enough rain nor warmth in the system to support good tree growth, and the area is prone to forest fires. The blueberry, a plant that can survive through long periods of flooding, drought, and fire, does well in the region. With increasing latitude northward in the taiga, the spruce forest becomes progressively sparse and the trees diminutive.

Even though the area is climatically defined as arid, there is considerable surface water. Evaporation is low, bedrock relatively close to the surface, and underlying permafrost prevents water from percolating very deeply. The small amounts of rain water tend to accumulate on the surface. In this fairly flat yet irregular landscape, water moves towards the sea, differentially confined by the topography. The difference between rivers and lakes in this region is often a descriptive difference of size and the rate at which the water is moving. Once constricted, lakes increase their current speed and become flowing rivers, only to broaden out again and turn into slow-moving lakes. Some lakes are parts of river systems, but there are myriad bog lakes. Bogs and beaver ponds have stable water levels and because of decomposition of vegetation, are often acidic and sterile compared to riverine lakes that fluctuate sometimes dramatically throughout the year.

In the taiga, aquatic environments (including coastal marine environments) are sometimes more productive than terrestrial environments. Additionally, large riverine lakes have greater diversity of species than small lakes, bogs, rivers, and estuaries. The reason stems in part from the slower movement of water, which allows soil and organic material to accumulate. The rest of the explanation lies in the peculiar nature of water. Unlike most other materials, water gets lighter (less dense) as it cools below four degrees Celsius. Water also stores considerable heat. Ice floats and insulates the unfrozen water below in deep lakes. Thus, large deep lakes store the heat of sunlight throughout the year, providing an environment where foraging and reproduction can occur year round. Small lakes get very warm during the summer but can freeze deeply in winter, whereas large lakes and reservoirs warm slowly throughout the summer and cool more slowly in winter. The result is that the larger reservoirs are "cold" in the summertime but on an annual basis are much "warmer" than small lakes. Shallow bays in the larger bodies of water, however, may act like the smaller lakes.

Rivers differ from lakes in that moving water has the energy to remove soil and organic material. The greatest volume and velocities are attained in the spring when melting ice and snow scour soils and plants from the river bottoms and surrounding countryside. The large amounts of ice (frazzle) in the roaring spring rivers add to the scouring and damage living tissue. In short, discharge extremes on rivers and their estuaries result in low total productivity and species diversity (Roy 1989).

The limiting physical characteristics of the taiga can be summarized as follows. Rivers contain too much water and ice in the spring. Small lakes, bogs, and beaver ponds often freeze more deeply in winter and can be acidic in summer. Dry land fosters little growth because it is too cold in the winter, which is most of the time, and too dry in the summer. The largest lakes have the greatest seasonal stability, high primary productivity, and the highest species diversity in the taiga ecozone. And large bodies of water are gradually going extinct in the area because of glacial rebound.

BIOLOGICAL CHARACTERISTICS OF THE JAMES BAY REGION
Questions about how plants or animals obtain carbon, nitrogen, or any other essential substance are paramount to understanding an ecozone. Many of the essential molecules are recycled within a system. Plants performing photosynthesis share the resulting sugar with specific bacteria, changing inorganic nitrogen into organic nitrogen, which the plant in turn uses to make protein. Animals eating the plants reorganize the plant protein into animal protein, and carnivores eating those herbivores again reorganize the protein to their own needs. Death begins decomposition cycles where scavengers or micro-organisms again rearrange the organic material into their own. They in turn grow, die, and become food for another organism.

Production of living cells is limited by the quantity of essential nutrients in the ecosystem. Organic nitrogen is one of those limiting compounds that accumulate in the system over time and are recycled over and over again, starting with decomposition of dead organisms. For example, microalgae living on the interface between sea ice and water depend on tidal movements to circulate organic nitrogen essential for growth (Demers et al. 1989). Increases in the growth of phyto- and zooplankton in lakes are usually a result of nutrients leaching into the lake. Accumulation is greater in lakes because erosive processes tend to be less severe than in other environments (Ploskey 1986).

Recognition of the leaching of nutrients from one habitat to another is also important in understanding how an ecozone functions. Productivity in lakes and rivers is a function not only of internal recycling but of leaching of nutrients from the surrounding landscape. Part of the reason why the land in this area is so relatively unproductive is that accumulated nutrients leach into aquatic, and eventually marine, environments. Most commercially harvestable saltwater fish, for instance, get their nutrients from terrestrial sources, which is why most marine fisheries are associated with large river systems that drain large land masses.

Therefore, in the cold taiga, primary photosynthetic production is low and decomposition is slow. Growth is highest in areas where accumulation of organic material is greatest, and these areas tend to be quiet aquatic environments.

Environmental Effects of Hydroelectric Generation on the James Bay Region

The argument has been made that giant reservoirs could have a serious impact on climate by affecting the circulation patterns of ocean currents, while also affecting various species by changing the annual freshwater regime. Mysak (1993) argued that the cumulative effect of extensive hydroelectric development in James Bay and Hudson Bay could change and cool the climates of Europe and other areas of the North Atlantic. At best these calculations, like those for the global warming hypothesis, are difficult to make because of the large number of variables to be considered. LeBlond, Lazier, and Weaver (1996) counter this argument, estimating that changes in the winter flow patterns of all the rivers in the two bays might increase winter ice formation by four percent – well within normal variation. When the metre of ice on James Bay melts in the spring, more fresh water is released into the sea than the annual discharge from all the rivers combined (Prinsenberg 1988). Similarly, when water freezes in the fall, the expelled salt raises the salinity and allows the water underneath to reach lower temperatures. The point is that the changes in seasonal freshwater flow are minuscule compared to other natural seasonal events.

Post-impoundment Canadian reservoirs initially increase in organic productivity. Flooding kills terrestrial plants and decomposition increases in the reservoir. Flooded peat bog mats float to the surface of the lake and drift around, decomposing gradually. The products of decomposition enter the food chain. After an initial bloom, which Verdon et al. (1991) estimate peaks in the first three years, productivity stabilizes at a level that is often slightly higher than the pre-impoundment level (Kimmel and Groeger 1986). Different reservoirs require varying lengths of time to reach a new equilibrium. In the Manicouagan reservoir system there is still a substantial amount of unrotted timber on the shores twenty-five years after flooding, and the seventy-year-old Gouin reservoir still has standing submerged trees. These older reservoirs are a testament to the slow rate of decomposition in Quebec. Leaves and soft organic material will decompose in less than a year while twigs may take up to three years.

Based on a comparison of older reservoirs (the Smallwood and Manicouagan reservoirs) with adjacent natural lakes, Verdon et al. (1991) estimated that it takes twenty to thirty years for mercury levels to reach pre-impoundment levels in piscivorous fish (primarily walleye and pike). This would also indicate that a new equilibrium in productivity has been reached. Already the levels of mercury in lake whitefish taken from the second La Grande reservoir have declined and are below the Canadian standards for human consumption of mercury in commercial fish. Equilibrium refers to the point at which pre-impoundment fertilizer (decomposable plant material that results in increased post-impoundment growth) runs out. The peak in mercury contamination was less than seven years, depending on which species and which reservoir was involved. The rivers that were diverted did not show increases in mercury. Some species, such as northern pike (*Esox lucius*), doubled in the levels mercury in their tissue, but often the mercury in this species exceeds Canadian standards even in natural lakes. (For a full discussion of the mercury issue, see chapter 4.)

It has been argued that reservoirs do not act like lakes and are not as productive as lakes because of fluctuations in water levels. Changes in water levels affect some species more than others, depending on a number of factors. Organisms in the water column are less affected by this phenomenon than those fixed to a surface.

Shoreline organisms face the greatest risks from receding water levels. The phenomenon known as the "ring around the Arctic" is this tidal line zone of rock, where algae and other plant life are scraped from the rock by the action of ice. This denuded ring is often wider in salt water because the tides raise and lower the ice, but the sterile ring is also wide in reservoirs because of the winter draw-down in water levels.

This drop of reservoir water/ice levels in the winter kills the eggs and young of species that breed in the shallows during the fall and hatch in the spring. Speckled trout (*Salvilinus fontinalis*) is an example of a species that is put at a disadvantage in northern reservoirs. The "natural" river/lake systems traditionally have widely varying water levels depending on season, storms, thaws, and many other climatic conditions. These natural bodies of water tend to maintain or increase in volume in the winter, whereas the Hydro-Québec reservoirs, which draw down the water to produce electricity for winter home heating, decrease in volume during this time of year. However, animals like northern pike, which breed in shallows in the spring, seem to do better in reservoirs than in natural lakes.

The decrease in the spring flush increases the productivity in the lower rivers, partly because of less scouring. In addition, rivers such as the Eastmain, with its greatly reduced flow, have a higher productivity than they had before the diversion. This is not only because the rivers are gentler on organisms but also because they are richer in nutrients. Prior to the diversion, the Eastmain waters were diluted with the sterile upland waters, whereas now they have a greater concentration of organic materials leaching from the low marshy coastal regions. The time that each molecule of organic material spends in the river is longer now because of the reduced flow; this means that growth and development can take place.

The situation is a little different on the Koksoak since the reduction of flow occurs further upstream; hence nutrients are leaching into the rivers at a higher rate. But here the reduced flow allows salt water to intrude further upstream. Since Ungava Bay is relatively rich in nutrients, the nutrients also increase in the lower Koksoak (Messier 1985, as reported by Harper 1992). In this case, because of the lower flow rates, higher nutrient concentrations, and higher salinity, the productivity has increased while the species composition changed.

The fluctuation in water levels in these northern reservoirs is very different than in reservoirs of similar size in the United States. The larger US dams on the Missouri, Colorado, and Columbia rivers not only produce electricity but play a major role in irrigation. This use requires draw-downs in summer as well as winter. It is just as difficult to generalize from reservoir to reservoir as it is from lake to lake.

Hydroelectric dams result in replacement of terrestrial and riverine species with lake-adapted forms. At these latitudes, with attendant low terrestrial productivity, most of the plant and animal species (as discussed above) are the marginal low-density populations from adjacent ecozones. In fact, the composition of river and lake species is not very different in the taiga.

Studies of other dammed rivers that flow (or no longer flow) into James Bay or Hudson Bay (e.g., the Moose, Mattagami, Churchill/ Nelson) present no evidence that the changes in seasonal flow or an increased freshwater plume have changed primary production in the salt water. What evidence there is indicates that primary production has increased; an increased nutrient flow has resulted from the increased decomposition available in the newly enlarged freshwater lakes.

Data on primary productivity in James Bay has shown no reductions in, or changes in species of composition of, primary producers. Harper (1992, 421) is worth quoting here: "James Bay is shallow (generally less than forty metres) and relatively unproductive due to cold water and low nitrogen (Messier 1995) and phosphorus (Legendre and Simard 1978) levels. Half of the discharge now comes between December and April, and the plume in the bay is increased in winter by a factor of three. There is little evidence that the productivity has changed, nor do the fisheries seem affected. The increased organic matter in the water does not cause much oxygen depletion because of the fast turnover of water."

The dynamics of the phytoplankton do not appear to have been affected greatly, although local and temporary effects due to changes in hydrodynamic patterns have been observed (Ingram et al. 1985).

On the same page, citing Julien and Laperle (1986), Harper reports that there has been an effect both on eelgrass, because of the freshwater plume, and on the waterfowl feeding on it. Citing Messier (1985), he claims that this in turn affects native hunting, but no particulars are given. The abstract of Messier et al. (1989) indicates that there has not been a very large change in the freshwater plume mainly because tidal rhythms seem more important in the mixing processes than total inflow. Lalumière et al. (1994) indicate no changes during their long-term studies of eelgrass productivity on either side of the La Grande estuary. We are not aware of data on native hunting success. The spring and fall waterfowl hunts are to collect food and have, in addition, ceremonial and ritualistc values that the Cree feel preclude monitoring processes.

Tracing the route of primary productivity, species diversity, and endangerment is one methodology used to assess the environmental impact of the hydroelectric facilities. We propose now to look at three classes of vertebrates – mammals, fishes, and migratory birds – and consider the effects of existing and proposed dams on the taiga shield. We will pick a few species that illustrate variations and concerns. We are interested not only in identifying environmental changes but in assessing the extent to which both resident and migrating wildlife are equipped to adjust to these changes.

MIGRATING AND RESIDENT MAMMALS

Mammal species decrease in diversity as latitude increases. The boreal forest just south of the taiga has thirty-six species of mammals. Just a few miles north, as one enters the taiga shield, the number drops by twenty-five percent to twenty-seven species. As mentioned, however, many of these species are circumpolar and regionally very populous.

One hypothesis is that terrestrial mammal diversity and numbers will diminish with the reduction in terrestrial habitats. There has been a four percent reduction in terrestrial habitat in the La Grande watershed. Fluctuations in size are normal and often cyclical in subarctic populations. Good estimates of past population numbers and their oscillations are not available, nor do we know how the geological processes will change these habitats in the future. But the slight increases in regional bioproductivity should produce population increases of terrestrial wildlife.

Population fluctuations in caribou illustrate the kinds of estimation problems involved. The La Grande project has flooded some land that was traditional grazing ground, but, as noted earlier, the nearly 200,000 caribou present in Quebec when the La Grande project began twenty-five years ago are now approaching one million animals. Wildlife managers fear a crash is imminent. Already animals are showing nutritional weaknesses in their bones – a sign that the population is overgrazing its range habitat. When crashes come, animals not only starve to death but die while migrating from winter ranges to the summer calving grounds. Often animals will be seasonally weak, which results in high mortality after the winter as animals try to cross rivers and get to summer feeding grounds. No methodology allows us to predict whether, if the reservoirs had not been created, the population might have gone to 1,250,000 animals before it crashed, or whether the population died back to below "normal" levels because of the dams.

The activity of beavers is another case in point. Beavers should not be underestimated for their effect on the landscape. Boreal and taiga habitats burn frequently and are flooded by beaver just as frequently. Flying over eastern Quebec, one is always staggered to see the beavers' industry in producing acres of dead and decaying forest. Beavers on the reservoir are one of the species affected by the winter draw-down of reservoir water. Lodges appear high and dry with their subsurface tunnels now above water. But in the Manicouagan reservoir – one of the biggest in Canada – beavers are flourishing.

Beluga whales (*Delphinapterus leucas*) are not very common in the estuary of the La Grande but are common in the river estuaries of Hudson Bay. This is a circumpolar species and currently quite numerous. Hunted by native people and others as a food source, it became regionally endangered in the middle of the last century. Specific herds reached very low numbers primarily because of overhunting. Canadian law forbids commercial harvesting, and the taking of animals by native people is regulated. In Quebec, the St Lawrence and Ungava Bay populations are classified as endangered. Reeves and Mitchell (1989, 235) suggest that the eastern Hudson Bay population of about 1,500 (740 to 1,940) animals should be classified as threatened because of "its depleted state, continued exploitation and exposure to environmental modification (particularly river damming and diversion)." Their objection to hydroelectric development stems from caution. They do not hypothesize a specific problem that the projects might cause for these animals, and they admit that not all changes would necessarily be harmful, "although it is not always clear whether the effects of development on a particular species will be favourable or unfavourable" (ibid., 234). They point out that the increase of cisco (*Coregonus artedii*) in the La Grand estuary might have been beneficial to the whales. On the other side of Hudson Bay, the Churchill River was redirected into the Nelson for hydroelectric generation, changing seasonal and regional flow patterns on a scale that is equivalent to the proposed Great Whale project. That area has over 20,000 belugas. Studies of the project could give insight into the effects of hydro dams on these marine mammals.

The beluga is a good illustration of how differently species are managed. There has been considerable alarm about the 1,500 beluga on the Eastmain. But the alarm is partly due to the population-monitoring system of the beluga managers. It is the regional herd that is thought perhaps to be endangered, not the species beluga, which at the present time probably has several hundred thousand individuals worldwide. The managers of belugas are not sure that a herd is even a biological unit of measurement. In other words, it is not known whether the 1,500 animals are a discrete unit, like a pack of wolves, or a few different individuals that roam out from the 100,000 animals using the arctic basins. So far nobody has offered evidence that the population of belugas in Hudson Bay is in any way endangered or bothered by the cumulative impacts of the hydroelectric developments around the bays.

There are two common species of seals in Hudson Bay, the ringed seal (*Phoca hispida*) and the harbour seal (*Phoca vitulina*). As with the whales, their populations were affected by hunting pressures. They are both tolerant of fresh water; some harbour seals actually take up residence in fresh water. Food supply and hunting pressures are the two main concerns for these species.

FISH

As with mammals, the number of fish species declines as one moves north in the region. The La Grande River drainage includes thirty-one species of fish in the western end; with increased latitude and elevation there are only sixteen in the Caniapiscau. On rivers coming into the southwest side of James Bay there are forty-four species (Harper 1992; Roy 1989). Growth is slow, as it is in most rivers this far north. Fish size has more to do with longevity than the fertility of the watershed. Harper (1992, 415) notes differences in growth between species: "Lake whitefish (*Coregonus clupeaformis*) and longnose sucker, or dace (*Catostomus catostomus*) and black suckers (*C. commersoni*) have near normal rates, while others (sturgeon [*Acipenser fulvescens*], goldeneye [*Hiodon alosoides*], sauger [*Stizostedion canadense*] and walleye [*S. vitreum*]) have very reduced growth." "Moreover, fecundity is low, sexual maturation retarded, and reproduction episodes in an individual may occur only at every second to sixth year. This leads to low population densities, often of the order of a few adult fish per hectare." Roy (1989) does not list goldeneye or sauger in the La Grande. Walleye exist only in the warmer bays on the south side of the second La Grande reservoir.

Walleye are a good example of the difficulty of judging the effects of impoundments. At the latitude of the La Grande they are at the limit of their range, although before impoundment they were the primary predator in the western La Grande region. In fact, as one travels east up the La Grande, the rise in altitude cools the waters to the point where walleye cannot reproduce. Thus, in the smaller pre-impoundment lakes, walleye were present because the lakes were warmer in the summer. But the second La Grande reservoir is so large that it does not get quite warm enough in the summer for walleye to reproduce successfully, except in the southern bays, and in the warmer waters of the Sakami Lake diversion where the warmer waters of the Opinaca empty into the reservoir. Walleye compete with northern pike, the walleye favouring areas of high turbidity. One of the other effects of the reservoirs is to increase the transparency of the water, which throws the advantage to the pike.

Fish populations were monitored from 1977 (a year before the second La Grande reservoir was impounded) to 1992, to test the effects of creating these large reservoirs. The methodology was to use gill nets and measure "catches per unit of effort." After the La Grande reservoirs were filled, seventeen species of fish were observed but only seven were regularly caught. Population *densities* declined during the "fill-up" (because of the larger volumes of water), but in three or four years they returned to pre-impoundment numbers, though with a noteworthy increase in biomass. Compared to large natural lakes in the region, the yields were the same but the fish were larger. Northern pike and whitefish were several times more abundant than in natural lakes. Even the small populations of cisco (*C. artedii*) and burbot (*Lota lota*) were twice as abundant as in natural lakes. Walleye and the sucker (*C. commersoni*) are now restricted to a few locations (Harper 1992).

DesLandes et al. (1995) report in detail on what appears to be the same set of data. They document their observations species by species and lake by lake. Species that favour larger, cooler, and cleaner lakes do better than species adapted to warmer, turbid waters.

Colonization is easier for species such as whitefish, which enjoy the increase in organic productivity and are not ill affected by the winter draw-down; trout, on the other hand, are affected by the increased competition with pike and by winter draw-down, which exposes the eggs they lay in the shallows.

Quebec is one of the premier places in the world for big speckled trout, which do not do well in reservoirs (or in large lakes, for that matter). The replacement of smaller lakes by large reservoirs replacing clearly diminishes habitat for this species. Similarly, lake trout (*Salvelinus namaycush*) appear to be doing poorly in the La Grande complex as a whole. During the fall both salmonids (the family name for trout, salmon, and char) lay eggs in the shallows that hatch in the spring. Because of the winter draw-down, these species have had reduced reproductive success in the reservoirs. Lake trout are also one of those species that "homes" to ancestral breeding sites. The creation of the reservoir now makes these sites inappropriate for successful breeding and it will take this species longer to recolonize the reservoirs. These two salmonid species need some clever management in the La Grande system.

Dams can block the reproductive migration of species that are strictly anadromous, that is, species that migrate from salt water to fresh water for reproductive purposes. In this region Atlantic salmon (*Salmo salar*) and Arctic char (*Salvelinus alpinus*) are found only in the rivers that feed into Ungava Bay and the Nastapoka River, which flows into Hudson Bay. The diversion of forty percent of the waters of the Caniapiscau into the La Grande system reduced the total volume of the Koksoak River, which flows into Ungava Bay. This allows the salt water to intrude farther up the river and has reduced the number of Arctic char by thirty percent, although the total yield of fish has been fairly constant (Harper 1992). There are no obligatory anadromous fish in the mouth of the Grande River. Species such as the lake white-fish, cisco, longnose dace, and brook trout (*S. fontinalis*) are found in brackish estuaries (Roy 1989). During the years following impound-ment, whitefish populations increased in several of the reservoirs, but in the thirty-five-kilometre La Grande estuary the cisco population has dropped probably because the water is now too cold. The dams spill from the bottom of the reservoir where the water is the coldest. Cisco differ from lake whitefish in that they prefer shallow lakes, which tend to be warmer in this area. They did not exist in the Caniapiscau before impoundment. Thus, at the margin of their northern distribution, they will not do as well as lake whitefish in the larger and deeper reservoirs.

Some fish migrate to spawn within a body of water. The problems of these diadromous fish (those migrating within fresh water for repro-duction) in hydroelectric development projects have been the subject of much study (Hall and Van Den Avyle 1986; Langford 1983). Two major issues are that the dams are barriers to upstream migration, and that changes in flow patterns confuse the spawner. In this context, it could be argued that the dams might locally endanger sturgeon. The dams on the La Grande River were located at natural barriers and fish access has not changed. The sturgeon in northern reservoirs is another example of a species that is at the extreme margin of a range that runs all the way to Alabama (MacKay 1963).

There have been several studies of the La Grande complex fish pop-ulations since about 1977. Most of them were done by Hydro-Québec personnel or in some way supported by the corporation and its affili-ated organizations. The criticism that they have an innate bias that would affect the way they report the results is probably true. But under Canadian law, the developer is responsible for doing environmental assessments, which are monitored by the provincial and federal gov-ernments. Many of the reports are also published in peer-reviewed journals, which should be impartial and judge the works on the merits of their methodologies.

The fact is that measuring changes that result directly from construction of the reservoirs is a complicated business. In the areas where the reservoirs were built, change is the rule. Just how reservoirs affect changes in the changes is difficult to reduce to single variables. Moreover, many of the changes that may have been induced by the reservoirs have not been realized yet. Reservoirs themselves mature over time. Beaches take years to form and stabilize, as do new underwater gravel beds that might be habitats for spawning fish. In the case of lake trout, the development of a suitable breeding habitat is also dependent on their finding it and re-establishing a "traditional" site. For many species of fish, colonizing these new reservoirs may take years.

Since there have been no measurable changes in the nutrients or plant life in the saltwater bay, one would expect few changes in fisheries. Cisco have increased slightly in the estuaries and brook trout have decreased. In these latitudes the marine food chain consists primarily of five trophic levels. Some species change trophic levels with age or season. Species such as the Arctic cod (*Boreogadus saida*) form important, even crucial, links between the various levels and are important indicators of the marine food web (Hobson and Welch 1992). Young age classes of fish, for example, consume plants and also often serve as prey for upper-trophic-level piscivorous fish of the same species. The Arctic cod is an example of a marine organism that may be affected by the decreased spring freshwater discharge. Gilbert et al. (1992) hypothesize that because these fish avoid the freshwater plume, larvae might be able to feed several weeks earlier than with a natural spring flood.

BIRDS

Some resident birds such as ptarmigan (*Lagopus lagopus*), have been displaced by the hydroelectric facility. Given the vastness of the upland areas and the sparseness of the population, it is hard to characterize this displacement as anything but minor. The problem of assessing the effect of hydroelectric facilities on populations of migratory bird species is more difficult. It is speculated that a number of migratory shorebirds and waterfowl might be adversely affected by the proposed projects, although few data are available.

Hydro-Québec claims the "displacement" of seven thousand nesting pairs of waterfowl at the La Grande complex. Displacement is not the best word to use because it assumes that the birds have somewhere else to resettle. Typically a species will fill its niches, and animals displaced by habitat loss might not be able to find an unoccupied suitable home. According to Austin Reed (pers. comm., 21 May 1993), most of the interior-nesting waterfowl are Canada geese (*Branta canadensis*) and black ducks (*Anas rubripes*). The number of Canada geese nesting in areas affected by the La Grande complex is relatively low (Bellrose 1976). The nesting population of black ducks in this area of eastern Canada is also quite sparse (Rusch et al. 1988).

A number of sources express concern about the harlequin duck (*Histrionicus histrionicus*). Rimmer's characterization (1992) is typical: the duck's breeding population "is classified as endangered and may number less than 1,000 birds. Some 200 Harlequin Ducks are thought to nest within the Great Whale project area." From this the reader might think that twenty percent of the species population would be put at risk if the Great Whale project were to go forward. But Rimmer is referring to the estimated population of the harlequin duck in eastern North America. Estimates of the harlequin across its entire range fall between 600,000 and 1,000,000 birds (Bellrose 1976). For the eastern population (the western North American population is much larger), the literature identifies Iceland as the major nesting area of this duck, with the eastern Atlantic shore from Maine north as its wintering ground (Bellrose 1976; Goudie 1989; Kortright 1953). None of these sources mentions James Bay, although Kortright's map (1953, 291) shows a peripheral population breeding on the rivers draining into Ungava Bay. Bellrose (1976, 381) notes that "in eastern Canada, Harlequins breed in sparse numbers over a wide area extending from southeast Baffin Island through Ungava Bay, Quebec, and coastal Labrador to the north shore of the St Lawrence and the Gaspé Peninsula."

Conversations with biologists in the region suggest that some harlequins may have nested on the Eastmain and La Grande drainage before the dams were built. Estimates place these birds at a few pairs, and most estimates indicate that more harlequins use the turbulent waters of the Great Whale River (Reed, pers. comm., 21 May 1993). It has been suggested that this eastern North American population of harlequin may be a distinct subspecies, which must be the basis of Rimmer's contention that twenty percent of the species population was endangered by Hydro-Québec. But this opens another debate, for there are ornithologists who do not believe the eastern and western populations are genetically different. The species and subspecies questions are always debatable among systemic biologists, mainly because the boundaries between subspecies are so ambiguous. It should be noted that Bellrose (1976) not only overlooks the possibility that the harlequin in eastern North America are a separate species but he also fails to observe the possibility of species differences between harlequins on the east and west coasts. He simply says that the harlequin's range "is divided into two separate and distinct regions: eastern and western" (Bellrose 1976, 381). For the record, similar divisions of range occur for the buffleheads (*Bucephala albeola*), goldeneyes (*B. clangula*) and other diving species, not to mention dabblers like mallards (*Anas platyrhynchos*), but there is no talk of subspecies with these ducks. This is one of the many cases where the taiga shield is a marginal habitat for an abundant circumpolar species.

Of greater concern is the long-term effect of cumulative impacts (Rimmer 1992) on James Bay itself. While the eastern side of James Bay is not a major nesting site for geese, its waters do provide important resting areas for migrating geese, particularly Canada geese, snow geese (*Chen caerulescens*), and Atlantic brant (*B. bernicla*). Bellrose (1976) notes flocks of Atlantic brant numbering as many as 60,000 birds gathering on the east side of James Bay. The Atlantic population of Canada geese, the largest North American flock, uses the east side of James Bay as an important resting area during its migrations.

In short, the east coast of James Bay represents an important stopping point for both Canada geese and Atlantic brant engaged in a 1,500-mile migration. "Resting or "stopping" points are popular terms for ecologically important concepts. It takes great energy to migrate long distances. Many species migrate in shorter intervals, stopping at feeding areas to refuel. The shores of James Bay are important for such species.

One concern is whether increased flow of fresh water through diversion will lower the salinity of the bay to the point where it will be frozen when migrating birds want to rest there. The La Grande, which accounts for thirty-four percent of the freshwater effluent in the bay (Harper 1992), has not had that effect. In fact, because of the increase of the winter discharge, the lower La Grande estuary no longer tends to freeze.

The lesser snow goose is another source of concern (Rimmer 1992). Recent research on snow geese examines the effect of nesting habitat on populations. Wildlife managers consider gosling mortality among snow geese in Manitoba to be density dependent. Moreover, while the North American Waterfowl Plan's Arctic Goose Joint Venture identifies James Bay hydroelectric construction as a key concern with regard to Canada geese and Atlantic brant, it fails to mention the project during discussions of the habitat needs and population status of the snow goose.

Upon closer look, it becomes clear that the hydroelectric facilities are of little concern because lesser snow geese make minimal use of the east coast of James Bay (Bellrose 1976). They prefer the vast intertidal areas on the south and west coasts of the bay. They do use the southern part of the bay extensively, in the estuaries of the Rupert and Broadback rivers. Any argument regarding the vulnerability of the lesser snow goose must be launched from the premise that the greatest danger they face is the long-term cumulative impact of all hydroelectric facilities in the James Bay region. To reiterate, to date there is no evidence of any such impact.

The migrant Canada goose population in the Atlantic flyway has dropped dramatically in the last few years. The nesting population in the Ungava region was estimated at three-quarters of a million birds in the late 1980s and a little over a quarter of a million birds in 1993. The problem is serious enough for some American states to have cancelled the hunting season on this species. At the present time it is not known why the decline has occurred. Some people think the spring goose hunt of the Cree is responsible, while others think Americans shoot too many birds. Meanwhile, there are ongoing complaints from golf courses, football fields, and parks about goose excrement from large populations of resident geese. Whatever the causes of these population shifts, so far nobody has linked the changes to the La Grande project.

With regards to raptorial birds, it is hypotheszed that methylated mercury moving up the food chain could reach concentrations high enough to poison adult birds and nestlings. The osprey population is currently increasing, in all likelihood because of the increase in fish around the reservoirs. However, there is currently no information on the effect of mercury on wildlife. Fish-eating birds such as loons and mergansers should be monitored.

Observed Problems of the James Bay Project

The real problem with impounding huge volumes of water in the taiga shield is not that a terrestrial environment is replaced by an aquatic one, or, from a more sophisticated perspective that a few big reservoirs replace hundreds of small lakes and streams. The James Bay impoundments do not exhibit the range of problems created by large US reservoirs – such as the Columbia, Missouri and Colorado systems – which include silting, evaporation, and salting. As argued above, such large bodies of water may actually increase regional biological growth.

The real problems lie, not in the consequences of the finished dams, but in the process of building them. Impounding large bodies of water entails massive construction. Huge amounts of gravel and rock must be moved for dams and dikes and hundreds of miles of roads and power lines brought into the region. If one applies to the construction process the same criteria applied to the finished impoundments (primary productivity, species diversity, and species endangerment), a different picture emerges.

Primary productivity in these areas is reduced. During construction, soil is removed or scarified. Primary productivity is altogether precluded by road surfaces, and unless replanted, roadsides and borrow pits can remain sterile for years. To a lesser extent the same tends to be true of power lines where heavy equipment damages vegetation, exposing soils to erosive processes.

In the taiga road construction means building the road bed well above the surrounding habitats and bridging water courses. In order to do this, vast quantities of gravel must be removed from borrow pits, natural deposits of glacial origin. These borrow pits, which may be several square miles in size, can remain virtual deserts for years. Hydro-Québec has planted much of the six-hundred-mile paved La Grande road and the five-hundred-mile gravel Caniapiscau road with nitrogen-fixing alder (*Alnus* spp.). Many of the borrow pits have been replanted but many others have not. No plantings have been done along the new road from Chibougamau to Nemiscau across the Eastmain River to join the La Grande road.

Power lines appear to do less damage to vegetation since the topsoil, for the most part, is not removed by construction. But because of damage to the plants and soil, erosion can remove the soil from sloping land, exposing bare rock, which will be unproductive essentially forever. In some places there is no soil to begin with, but trees and lichens provide a vegetative cover that is easily disturbed and disappears during construction.

The decrease in primary productivity in construction areas will, of course, reduce the number of resident organisms. Some ecologists claim that "edges," or boundaries bewteen two zones such as a power corridor and a forest, are beneficial to species diversity, but if the lack of primary productivity is what is defining the edge, then the argument is negated. The environmental impact of the roads on water courses seems to have been minimal, attesting to a measure of care and regard for the environment during construction; one does not observe, for example, the acres of drowned forest that are so typical on New England highways. But the number of culverts and bridges that cross streams, rivers, marshes, and bogs is enormous.

Perhaps the most serious problem is that roads expose wildlife to deleterious interaction with humans. The road to Radisson attracts over twelve thousand visitors a year. The road from Chibougamau has allowed a flood of campers, hunters, and fishermen into an area that was previously inaccessible. Corridor hunting and fishing leave an increasing number of lakes and marshes vulnerable to overexploitation. Wolves and other top-level carnivores are particularly vulnerable to road density.

The total effect of roads is infinitely complicated (and may, in fact, be the most serious threat to wildlife worldwide). Roads form the leading edge of increased human penetration and impact. In Quebec the dam-oriented roads have opened regions to a number of human uses and resources, especially mining and forestry. They have improved the economics of logging for paper production. The problems of Canadian timber policies are enormous and could scarcely be discussed in this chapter. Suffice it to say that the popular belief that trees are a renewable resource is probably not true: removing trees from the region transfers nutrients out of the area, scarifies and ultimately damages soils, and thus limits the growth of each succeeding generation of trees.

The roads also provide market opportunities for native people. Polyethylene gill nets are currently being used to procure fish for sale. There are quotas on how many fish can be sold from each lake, but the limited data on which the quota decisions are based are being collected retrospectively. Harvestable growth of walleye in the regional lakes will range from about 0.4 kilogram per hectare per year in the South to 0.1 kilogram per hectare per year in the North. With growth levels that low, it is very easy to make serious mistakes.

Sturgeon is a species at the margin of its range (at the La Grande), and it takes twenty-five years to reproduce. Not surprisingly, the population is quite vulnerable to netting. There is no "catch and release" policy for gill-netted fish; once caught, they quickly die. Thus, it is difficult, at best, to manage this resource by limiting the species that might or might not be netted. What of the "extra" sturgeon incidentally taken during the legal netting of pike, trout, and other species?

The roads also allow access for people with sophisticated equipment. Boats with fish locators and outboard motors can be transported by all-terrain vehicle to lakes that until a few years ago were accessible only by primitive techniques, or to the few who could afford to fly in. These high-tech encroachments are taking place in a province with a high percentage of people who regard hunting and fishing as an inalienable right.

Subsistence harvesting by native peoples, which *is* for them an inalienable right, has to be balanced against modern demographics. Historically the subsistence population of James Bay Cree may have been less than 1,500 people. As noted elsewhere, that population is now rapidly approaching 14,000. At the present growth rate, there could be 30,000 Cree in the James Bay region by 2020. While the absolute number of full-time Cree hunters and trappers under the income security program has remained stable, such activities as the spring goose hunt, corridor hunting along roads, moose and bear hunting, and fish netting increase as population grows. Canadians are coming to realize that the harvesting of wildlife by native peoples is significant and needs to be managed more closely (Berkes 1990; Reed 1991).

Thus, the increasingly articulated network of roads, combined with growing population both without and within the territory, serve to facilitate and intensify the pressure on this ecozone.

Conclusions

Most of the species of birds and mammals we observe today evolved some three to five million years ago. All the species using the James Bay region are much older than James Bay itself. Many of these species have seen James Bay disappear for thousands of years, reappear as a huge freshwater lake, and only recently become a saltwater inlet. In the next few thousand years James Bay will disappear again, transforming into a terrestrial fen as glacial rebounding dumps the water into Hudson Bay.

The taiga shield lacks the productivity to support a wide diversity of year-round populations. The migrating animals (both birds and mammals) that frequent the James Bay region take advantage of seasonal blooms. They capture the energy from several ecozones, rather than adapt to a specific zone the way an endemic resident species would. Migrant waterfowl capture the bloom of the tundra, when the days are twenty-four hours long, then move to the taiga when the light begins to fail, feeding on the summer's produce; then they fly on to places like Chesapeake Bay for the winter. The absence of large resident populations to make use of these limited seasonal blooms provides an opportunity for migratory species, which arrive for a short period, collect the season's growth, and move away with it.

Migrating animals are adapted to moving from one zone to another. But this creates a population problem. We would like to think of a population that is in balance with its environment. But migratory species have to have populations that are in balance with each of the zones they visit. They may fly from a lush summer into a winter wasteland. The population of migratory caribou will eventually crash and thousands of animals will weaken and die as their burgeoning numbers overeat one or more of their seasonal food supplies.

This is a chronic problem for many northern species. Should we manage caribou to counter this age-old problem? At the present time the number of Canada geese in the Atlantic flyway is so low that the US Fish and Wildlife Service has closed the hunting season. Why is the population so low? Is it one of these cyclic events? Is it habitat destruction on the wintering range? Overhunting by Americans? By the Cree? Mismanagement on nesting grounds by the Canadians? Can cyclic events really be caused just by two cold nesting seasons back to back, along with some mismanagement, or are they cumulative effects?

Cumulative effects are serious considerations, too important to be wielded without evidence. The problem with the taiga is that populations of plants and animals oscillate considerably from year to year, and the oscillation is normal. In a sense, environmental impact statements must hit a moving target. The essential question in assessing development projects is to ask how the resulting change is going to affect the environmental changes occurring "naturally." We need an environmental calculus asking for the solution to the changes to change.

In the case of James Bay there is no reason to believe that the existing reservoir systems of the La Grande complex will in themselves initiate the ecological disaster that some predict; for one thing, the region's geological changes are inherently far more dramatic than those caused by the hydroelectric reservoirs. Looking at the lake-rich map of Canada, it is hard to imagine that adding or subtracting other lakes, even with their aberrant winter discharges, is going to cause an environmental disaster.

But such a statement should not be interpreted to mean that the entire province could be re-engineered without environmental consequences. Most of the river systems in the region run east to west and are roughly confined to a single ecozone. Changes in flow patterns and heat budgets tend to be latitudinally conservative. However, once one begins diverting large rivers north and south, then perhaps another analysis would be appropriate.

The diversion of the Eastmain River north into Lake Sakami and then into the second La Grande reservoir is a case in point. The Eastmain-Sakami diversion brings warmer southern waters to the reservoir – enough warm water to allow walleye reproduction to take place in and around the mouth of the Sakami River. To some extent this compensates for the reduction of walleye reproduction that had occurred when the small, seasonally warmer lakes were replaced by the huge La Grande second reservoir. Similarly, the Caniapiscau River naturally flowed north, and some large percentage of that river was diverted west into the La Grande system. This should have the effect of cooling the Caniapiscau, if only slightly, where it enters Ungava Bay. We do not know what the ecological effects of that have been, but even slight changes of temperature can have large effects. A temperature change of one degree can transform water from a liquid to a solid.

The recently proposed diversion of the Rupert River from the boreal forest ecozone north into the La Grande would take considerably more heat north into the taiga. Diverting the Great Whale watershed south would bring cooler water south, essentially piping tundra temperatures to the taiga. Temperature is the main consideration in productivity, species diversity, and endangerment in these regions.

Our analysis has focused primarily on the taiga ecology simply because construction of the La Grande complex was largely limited to the taiga. At the time we started our study there were long range-plans to build reservoir complexes on the Great Whale drainage and also on the Rupert, Nottaway, and Broadback rivers. These plans, like the La Grande project, were roughly constrained within a latitudinal ecozone. But the proposals have recently been modified. The current plan is to mix the waters of the taiga with colder tundra rivers and the warmer boreal forest waters – in other words, to run several additional rivers through the Grande turbines. To analyze this proposal would require a study of the ecological consequences of the heat transfers among the taiga, tundra, and boreal regions. The consideration thus becomes more complicated.

As we have pointed out, the La Grande complex is not going to be an ecological disaster. But the roads leading to it and the increasing human population could be. People can have a very real impact on productivity, diversity, and species survival. Seasonal human immigrants and permanent residents are not dependent on the region's photosynthetic output for their survival, as the regional wildlife is. People's food, medicine, clothing, fuel, and shelter come mostly from outside the region. The increasing number of people, be they miners, foresters, developers, electric workers, vacationers, campers, hunters, anglers, or residents, all change the nature of the region. Habitat loss through construction and development reduces the region's bioproductivity. It is the increasing stresses placed on individual species because of their commercial, ritual, or recreational value that further reduce productivity and diversity.

Once upon a time the "balance of nature" ruled and population blooms and crashes went unnoticed except by the few hardy people who could get that far north on foot or by boat. Now visitors can drive to the taiga at a hundred kilometres an hour on a paved road. The region has lost its wilderness status. From now on the residents of Quebec will have to monitor and manage the James Bay region very carefully to preserve its bioproductivity and diversity, so as not to endanger any species.

In short, it is becoming increasingly inadequate to speak of "our place in nature." The imperative to manage the global village in order to save it is leading not to the discovery of a new place for humans in wild nature but to a radically different understanding of the character of Nature in a domesticated world (Warner et al. 1996).

REFERENCES

Bellrose, Frank C. 1976. *Ducks, Geese, and Swans of North America.* Harrisburg, PA: Stackpole Books.

Berkes, F. 1990. "Native Subsistence Fisheries: A Synthesis of Harvest Studies in Canada." *Arctic* 43:35–42.

Chiras, D.D. 1985. *Environmental Science.* Menlo Park, CA: Benjamin/Cummings Publishing.

Demers, S., L. Legendre, S.Y. Maestrini, M. Rochet, and R.G. Ingram. 1989. "Nitrogenous Nutrition of Sea-ice Microalgae." *Polar Biology* 9:377–83.

DesLandes, J.-C., S. Guénette, Y. Prairie, D. Roy, R. Verdon, and R. Fortin. 1995. "Changes in Fish Populations Affected by the Construction of the La Grande Complex (Phase 1), James Bay Region, Quebec." *Canadian Journal of Zoology* 73:1,860–77.

Dudley, J.P. 1992. "Rejoinder to Rohlf and O'Connell: Biodiversity as a Regulatory Criterion." *Conservation Biology* 6:587–8.

Ford, T.E, and R.J. Naiman. 1988. "Alteration of Carbon Cycling by Beaver: Methane Evasion Rates from the Boreal Forest Streams and Rivers." *Canadian Journal of Zoology* 66:529–33.

Gilbert, M., L. Fortier, D. Ponton, and R. Drolet. 1992. "Feeding Ecology of Marine Fish Larvae across the Great Whale River Plume in Seasonally Ice-Covered Southeastern Hudson Bay." *Marine Ecology Program Series* 84:19–30.

Goudie, R.I. 1989. "Historical Status of Harlequin Ducks Wintering in Eastern North America." *Wilson Bull* 101:112–14.

Hall, G.E., and M. Van Den Avyle, eds. 1986. *Reservoir Fisheries Management.* Bethesda, MD: Reservoir Committee Southern Division, American Fisheries Society.

Harper, P.P. 1992. "La Grande Rivière: A Subarctic River and a Hydroelectric Megaproject." In *The Rivers Handbook: Hydrological and Ecological Principles,* edited by P. Calow and G.E. Petts: Blackwell Scientific Publications.

Hobson, K.A., and H.E. Welch. 1992. "Determination of Trophic Relationships within an Arctic Marine Food Web Using Delta-13 c and Delta-15 N analysis." *Marine Ecology Program Series* 84:9–18

Kimmel, B.L., and A.W. Groeger. 1986. "Limnological and Ecological Changes Associated with Reservoir Aging." In *Reservoir Fisheries Management,* edited by

G.E. Hall and M. Van Den Avyle. Bethesda, MD: Reservoir Committee Southern Division, American Fisheries Society.

Kortright, F.H. 1953. *The Ducks, Geese and Swans of North America.* Harrisburg, PA: Stackpole Company and Wildlife Management Institute.

Langford, T.E. 1983. *Electricity Generation and the Ecology of Natural Waters.* Liverpool, UK: University Press.

Lalumière, R., D. Messier, J.-J. Fournier, and C.P. McRoy. 1994. "Eelgrass Meadows in a Low Arctic Environment, the Northeast Coast of James Bay, Quebec." *Aquatic Botany* 47:303–15

LeBlond, P.H., J.R. Lazier, and A.J. Weaver. 1996. "Can Regulation of Fresh Water Runoff in Hudson Bay Affect the Climate of the North Atlantic?" *Arctic* 49, no.4:348–55.

MacKay, H.H. 1963. *Fishes of Ontario.* Department of Lands and Forests. Toronto: Bryant Press.

Messier, D., S. Lepage, and S. De Margerie. 1989. "Influence du couvert de glace sur l'étendue du panache de La Grande Rivière (baie James)." *Arctic* 42:278–84

Mysak, L.A. 1993. "Climate Variability and Change with Respect to Hydroelectric Development in Northern Quebec." Montreal: Great Whale Public Review Support Office.

Naiman, R.J., J.M. Melillo, and J.E. Hobbie. 1986. "Ecosystem Alteration of Boreal Streams by Beaver (*Castor canadensis*)." *Ecology* 67:1,254–69.

Pielou, E.C. 1992. *After the Ice Age.* Chicago: University of Chicago Press.

Ploskey, G.R. 1986. "Effects of Water-Level Changes on Reservoir Ecosystems, with Implications for Fisheries Management." In *Reservoir Fisheries Management,* edited by G.E. Hall and M. Van Den Avyle. Bethesda, MD: Reservoir Committee Southern Division, American Fisheries Society.

Prinsenberg, S.J. 1988. "Ice-cover and Ice-Ridge Contributions to the Freshwater Contents of Hudson Bay and Foxe Basin." *Arctic* 41:6–11.

Reed, A. 1991. "Subsistence Harvesting of Waterfowl in Northern Quebec: Goose Hunting and the James Bay Cree." *Transactions of the Fifth-Sixth North American Wildlife and Natural Resources Conference,* Washington DC.

Reeves, R.R. and E. Mitchell. 1989. "Status of White Whale, *Delphinapterus leucas,* in Ungava and Eastern Hudson Bay." *Canadian Field Naturalist* 10:220–39.

Rimmer, C.C. 1992. "James Bay: Birds at Risk." *American Birds* 46:216–19.

Roy, D. 1989. "Physical and Biological Factors Affecting the Distribution and abundance of Fishes in Rivers Flowing into James Bay and Hudson Bay." *Proceedings of the International Large River Symposium.* Canadian Special Publication of Fisheries and Aquatic Sciences. Publication # 106.

Rusch, D.H., D.D. Ankney, H. Boyd, J.R. Longcore, F. Montalbano III, J.K. Ringelman, and V.D. Stotts. 1988. "Report to the Wildlife Society on Review of Population Ecology and Harvest of the Black Duck by the Ad Hoc Tech-

nical Advisory Committee on Black Duck Conservation and Management."
Bethesda, MD: The Wildlife Society.

Verdon, R., D. Brouard, C. Demers, R. Lalumière, M. Laperle, and R.
Schetagne. 1991. "Mercury Evolution (1978–1988) in Fishes of the La
Grande Hydroelectric Complex, Quebec, Canada." *Water, Air, and Soil Pollution* 56:405–17.

Warner, S., M. Feinstein, R. Coppinger, and E. Clemence. 1996. "Global
Population Growth and the Demise of Nature." *Environmental Values*
5:285–301.

Wilson, E.O. 1992. *The Diversity of Life.* New York: W.W. Norton.

4 Elevated Mercury in Fish as a Result of the James Bay Hydroelectric Development: Perception and Reality

B.D. ROEBUCK

Introduction

In the middle of the 1980s, rumours of high levels of mercury in fish from the reservoirs of the Hydro-Québec development in the James Bay region started coming to public attention through brief newspaper, radio, and television accounts. By the early 1990s slightly more detailed reports were appearing in popular and well-respected magazines: *Canadian Geographic* (Gorrie 1990), the *New York Times Magazine* (Verhovek 1992), and *National Geographic* (Mitchell 1993). Unfortunately, these accounts dealt superficially with the mercury problem and were of virtually no value in the assessment of the seriousness of elevated mercury levels in fish caused by the hydroelectric project.

As a frequent visitor among the people of the Far North, and as an experimental nutritionist and toxicologist with extensive experience of various food-borne toxicants (though not mercury), I found these reports disturbing. Mercury has a long history as a toxicant (Clarkson 1991; Hamilton and Hardy 1974; Marsh 1987; Ramade 1987; World Health Organization 1990). In its organic form as methyl mercury, it is known to enter the aquatic food chain and bioaccumulate, increasing its concentration as it passes up the food chain to predator fish. Since fish are an important and highly nutritious component in the diet of many James Bay Cree, these reports suggested the possibility of a serious health hazard.

To explore the nature and impact of the mercury threat, I made several additional visits to the area and have met with Cree leaders, members of the James Bay Mercury Committee, and various scientists who have been working on the problem, most of whom are in the employ of Hydro-Québec. Although there is a substantial scientific peer-reviewed literature on the chemistry and toxicology of methyl mercury, most of the considerable research into the specific concerns of the James Bay mercury problem has been carried out by scientists contracted or directly employed by Hydro-Québec, and their reports, while of consistently high quality, have not been subject to peer review. Many of them have been published as a series of documents sponsored by the James Bay Mercury Committee, a joint Hydro-Québec/Cree effort established in 1986 (James Bay Mercury Committee 1994–95). What emerges is a picture that begs as many scientific questions as it answers regarding the impact of mercury on the health of the Cree, but it is also a picture that leaves no doubt regarding the psychological and social impact on the Cree of the mercury threat.

The maps on pages xvii, xviii, and xix show the specific regions of the James Bay region. The exact location of the reservoirs is particularly important since flooding of the reservoirs has occurred in stages and the concentrations of mercury in particular fish species vary with the age of the reservoir. Additionally, the Cree are affected to a greater or lesser extent depending on the proximity of their communities to the modified fishing areas (reservoirs and rivers). Finally, the recommendations regarding fish consumption focus not only upon species and their size but also upon where they are harvested.

Mercury Compounds and Their Biological Effects

Mercury has a long and colourful history as an environmental toxicant (Hamilton and Hardy 1974). Ancient Greek and Roman accounts report poisoning from the mining and smelting of mercury ores. In the early to middle 1600s, mercury nitrate entered the hatters' trade through the manufacture of felt hats and mercury poisoning has been associated with this industry up to recent decades (the "Mad Hatter" was exhibiting neurological symptoms of mercury poisoning). Numerous mercury compounds are still being used as pesticides, especially as coatings on seeds to prevent rot prior to sprouting (Hamilton and Hardy 1974; Ramade 1987).

Mercury exists in three chemical forms: elemental, inorganic, and organic. Mercury in thermometers and electrical switches is elemental, or metallic, mercury. In this form it is a liquid at room temperature and quite volatile. The vapour is toxic, but acute and chronic toxicity from elemental mercury is mainly an occupational risk. However, the vapour can be transported over long distances, ultimately re-entering the biosphere where it can be changed chemically into more toxic forms.

Mercury also exists in the form of non-volatile inorganic salts, such as the mercury nitrate that poisoned the hatters. The toxicity of these salts is variable; some, such as mercury nitrate, are highly toxic while others, such as calomel (mercuric chloride), have been used for medical purposes. Generally, mercury in geological formations and deep soils takes the form of inorganic salts.

When mercury combines with carbon containing molecules, organic forms of mercury result. For our purposes the most important mechanism in this transformation is the action of bacteria on soils and plant material that contain mercury, specifically decaying trees, bushes, and plant matter in flooded reservoirs. One common form of bacteria-generated organic mercury is methyl mercury. Methyl mercury formed by bacterial action is then incorporated into algae and other plants that can be eaten by herbivores, including inverte-brates and bottom-feeding fish. These organisms can in turn be eaten by predator fish, other animals, or humans. The methyl mercury is readily and efficiently absorbed from food by animals, and since it is preferentially soluble in fatty materials, it is poorly excreted and tends to accumulate in the flesh of the fish and to increase in concentration as the fish grow and continue eating mercury-laden food. In this sense, it is similar to the insecticide DDT and common organic pollutants such as the PCBS.

Like DDT and PCBs, methyl mercury concentrations increase as they are passed up the food chain, through a process called bioaccumula-tion. Fish that primarily eat plant materials are eating a food that is rel-atively low in methyl mercury, whereas fish that eat other fish are ingesting food in which mercury has already accumulated from lower trophic levels. If by culture or other circumstances the human diet is primarily composed of such fish, then the potential for ingesting too much mercury increases, and of course the ingestion rate will be higher if the diet consists of predator fish rather than fish lower on the food chain.

The Historical Basis of Methyl Mercury Intoxication

The first detailed account of methyl mercury poisoning was provided by Hunter et al. (1940), who reported on four men who were exposed to methyl mercury in the manufacturing of seed fungicides. Symptoms of poisoning began three to four months after exposure to methyl mercury had begun and continued to develop after exposure had ceased. The first symptom described by these workers was parethesia of the extremities. Parethesia is the sensation of burning or prickling similar to what is commonly described when blood flow is constricted and the foot or hand "goes to sleep." This was followed in about a month by ataxia of gait, imperfect articulation of speech, and impaired vision. These symptoms progressed even after exposure to the methyl mercury had ceased. These symptoms and their characteristic sequence of development have been termed Hunter-Russell Syndrome and are typical of methyl mercury poisoning. Subsequent research has shown that there is enhanced risk of severe and irreversible toxicity in the young, especially from foetal exposures.

The first recorded epidemic of methyl mercury poisoning occurred in Japan (Marsh 1987). Between April and December 1956, fifty people came down with a strange disease in the rural fishing area and village of Minamata. Initially, the cause was unknown; the likeliest possibilities were thought to be infectious disease, nutritional deficiencies, agents of abuse, or genetic factors. Eventually, the characteristics of the disease at Minamata were shown to be similiar to Hunter-Russell syndrome and the source was traced to industrial contamination of the fishing grounds with mercury. Industrial mercury waste discharged into Minamata Bay was converted biologically to methyl mercury, whereupon it concentrated in the food chain. In 1965 a second Japanese outbreak of methyl mercury poisoning occurred in Niigata, resulting in over five hundred recognized cases within a decade. There have been approximately two thousand cases of Minamata disease in Japan.

Minamata disease, as methyl mercury poisoning is now often termed, is remarkable from several points of view (Marsh 1987). In adults, a wide range of symptoms occur, the most common initial symptom being paresthesia of the extremities, sometimes accompanied or closely followed by constricted visual shields, ataxia of limbs and gait, slow slurred speech, deafness, and diminution of the sense of touch. Cases of mercury toxicity in newborns were reported at Minamata. Generally, the mothers were asymptomatic during pregnancy, though there have been cases in which the mothers had paresthesia. Severe psychomotor retardation in the children (somewhat similiar to cerebral palsy) was characterized by failure to crawl, stand, and speak recognizable words. In these cases, there was a general failure to meet age-specific developmental stages. There was some improvement with age, but almost half of the foetal cases were still severely affected a decade later. Clearly, foetuses are more severely affected by methyl mercury than adults. In Minamata there were many cases of foetal toxicity, while in Niigata only one foetal case was identified; there is no obvious explanation for the differing incidence of foetal disease.

Another extensive episode of methyl mercury poisoning occurred in Iraq in the winter of 1971–72 (Marsh 1987) Seeds intended for planting are still treated with fungicides, typically methyl mercury, to prevent rotting in the ground before germination. Seeds so treated are commonly dyed to indicate that they are not intended for consumption by livestock or humans. Wheat seed treated with methyl mercury was provided to farmers, who washed out the red dye, fed the seed to livestock, and, noting no apparent effects, began to consume the seed themselves. From December 1971 to March 1972, approximately 6,500 patients were admitted to hospitals and nearly five hundred died. This unfortunate epidemic of poisoning in Iraq afforded important data regarding the neonatal effects of methyl mercury. About ninety mother-infant pairs were studied. Estimates of methyl mercury exposure were made from the analyses of mercury deposited in the maternal hair at the time pregnancy. In all cases maternal symptoms during pregnancy were mild and reversible, paresthesia being the most frequent symptom. Foetal outcomes ranged from mild to severe symptoms of methyl mercury poisoning. Clearly, the foetus was far more susceptible than the adult. It was also noted in the Iraqi outbreak that minor time delays in child development could occur without permanent consequences.

Toxicity and Public Health Guidelines for Mercury Ingestion

The estimation of how much methyl mercury can be safely ingested without adverse effects is complicated and open to debate. Mercury concentration in the bloodstream is probably the best fundamental measure of risk, but this concentration is a balance between consumption and excretion, and that probably varies with age, physical condition, and other factors. Nevertheless, standards must be established, and on the basis of data gleaned from actual poisonings it is estimated that the lowest blood levels at which methyl mercury poisoning occurred was two hundred parts per billion. One part per billion is approximately one millionth of a gram of methyl mercury per litre of blood. Because of uncertainties, a safety factor is applied to ensure that sensitive individuals will be protected. In the case of methyl mercury, the values were divided by a factor of ten, leading to an estimated "safe" level of twenty parts per billion of methyl mercury in the blood. Since mercury is concentrated and permanently deposited in hair as it grows, the body burden of mercury is often estimated by measuring the mercury in hair samples. It has been determined that the hair concentration is about 250 times that of the blood, so that a blood concentration of twenty parts per billion of mercury corresponds to a hair concentration of approximately five parts per million, that is five millionths of a gram. Hair sampling is not only much easier than blood sampling but it also provides a historical record of past mercury ingestion.

Translating these estimates of a safe body burden of methyl mercury into a guideline for fish consumption obviously involves further assumptions regarding the rate of ingestion and excretion, as well as the size and age of the individual. Observations of populations that consume large quantities of fish indicate that the threshold for onset of symptoms (two hundred parts per billion of mercury in blood) corresponds to a long-term daily intake of 300 millionth of a gram of mercury for a seventy-kilogram (155-pound) person. Applying the safety factor of ten gives an acceptable ingestion rate of 30 µg of mercury in the diet per day. This value is termed the Acceptable Daily Intake (ADI), or, as it is now called Reference Dose Daily (RfD). Taking the next step of establishing a safety level of mercury in fish requires an estimate of the amount of fish that will be eaten per day, and this varies widely. The World Health Organization, for example, found that the average daily consumption rates for fish in Japan, Sweden, and the United States were 84, 56 and 17 grams. Based on those numbers, the safe concentration of mercury in fish would be 0.5, 0.7, and 2.4 parts per million respectively. Canada has adopted the conservative value of 0.5 part per million and the United States 1.0 part per million for the mercury content of fish tissue.

Before applying generalized guidelines such as the ADI, or the "safe" level of mercury in fish, to a specific population such as the Cree, it is important to remember the uncertainties and assumptions that were involved in reaching them. The ADI provides a guideline for the quantity of contaminant that can be ingested by an average person over a very long time without adverse effects, and it includes a safety factor such that the vast majority of the public would be protected. There is still some concern for populations that habitually consume large quantities of fish, or fish that contains relatively high mercury concentrations, or both, as may be the case with the Cree. Studies of selected populations that traditionally have consumed large quantities of fish have failed to observe adverse effects in either adults or children (Egeland and Middough 1997; Tollefson and Cordle 1986). In some of these populations, fish was consumed one or more times per day, year after year. Ideally, estimates of mercury intake by the Cree would incorporate details regarding the species and size of fishes eaten, the source of the fish, and the amounts of flesh consumed each day over a long period. Unfortunately, fish are eaten sporadically and the types of fish vary with season and location, so that the rates of mercury ingestion vary greatly. One response to such variation is to try to identify and protect whichever groups are at greater risk. This is exactly the approach taken by the James Bay Mercury Committee.

Mercury levels in either blood or hair have been used to monitor exposure to mercury, and attempts have been made under the auspices of the James Bay Mercury Committee to advise the population as a whole of its mercury levels. More specifically, attempts have been made to identify and monitor the people who are at greatest risk, such as those who consume large quantities of fish , as well as children and women of reproductive age who may be particularly sensitive to mercury. Older people and hunters and trappers who live a "traditional" bush lifestyle are typically large consumers of fish during at least part of each year.

History of Mercury Contamination in Northern Canada

The presence of mercury in the aquatic environment in Canada has been extensively documented (National Research Council Canada 1979). Several processes resulting in increases in environmental mercury have occurred more or less simultaneously and considerable confusion still remains as to the absolute importance of each process. There has been serious industrial contamination of geographically limited areas. Metallic mercury was used in an electrolytic process to produce chlorine and sodium hydroxide. The chlorine was ultimately used for the bleaching of paper. Pulp and paper plants in northern Canada used large quantities of mercury for such purposes and released metallic mercury into the aquatic environment. Mercury has also been associated with long-range atmospheric transport from centres of industrial activity, especially smelters. Additionally, mercury is present in fossil fuels such as coal and is released upon burning. There is some evidence that the phenomenon of acid rain, associated with coal combustion, and the acidification of lakes makes mercury more available to aquatic organisms, including fish (Xun et al. 1987). Two important but less well defined sources of mercury are vaporization from volcanic activity and from soils. Metallic mercury is volatile at ambient temperatures and it is widely distributed by atmospheric transport, so that mercury is found in relatively "pristine" environments such as the Greenland ice sheet (Weiss et al. 1975).

Well before construction got underway, all those involved in the hydroelectric development at James Bay were concerned by the widely publicized mercury poisonings in Japan, identification of mercury in fish in northern Canada, and suspected poisonings of native peoples. During the 1970s fifteen studies were published in Canada regarding mercury ingestion by humans, most of them looking at native populations (National Research Council Canada 1979). Two of these studies were of critical importance in that they heightened the concerns of the Cree near the hydroelectric development in the James Bay region. The first showed that natives living on or near the White Dog and Grassy Narrows reserves in northwestern Ontario had high concentrations of mercury in blood and hair. When examined, about ninety volunteers from these two communities exhibited one or more of the neurological symptoms characteristic of methyl mercury poisoning. The second study identified two probable cases and five definite cases of methyl mercury intoxication on preliminary examination of native people from the Matagami area of northwestern Quebec. Additionally, media coverage during the 1970s indicated that the Cree of northwestern Quebec consistently had higher reported blood levels of mercury than any other native population in Canada. Uncertainty contributed to official and public anxiety. The identification of what looked like Minamata Disease in domestic cats on Indian reserves in northern Ontario increased the parallel with the situation in Minamata Bay, Japan, where cats had also been poisoned (Takeuchi et al. 1977). But Wheatley et al. (1979) clearly point out that diagnosing methyl mercury poisoning with certainty is exceedingly difficult.

The marked elevation of mercury that has been recorded in many new reservoirs worldwide was a surprise. Bodaly et al. (1984) were among the first to document the problem with new reservoirs clearly. In northern Manitoba, concentrations of mercury in fish increased above natural mercury concentrations within two to three years in newly flooded reservoirs. These increases were highest in those reservoirs whose flooding submerged the greatest land surfaces. Bodaly et al. (1984) found that the total quantity of organic material submerged in the reservoir was the immediate source of mercury to the fish. The sediments were of lesser importance in supplying mercury.

Against the tragic background of Minamata, the identification of mercury in the environment and its movement into fish and ultimately into human beings produced considerable fear. As we have seen, actual assessment of the risk requires detailed information regarding mercury concentrations in fish, as well as rates of human ingestion of fish. The values vary greatly, but extensive surveys of types and amounts of food ingested *vis-à-vis* mercury content (see National Research Council Canada 1979; Tollefson and Cordle 1986) are periodically updated and repeated in Canada and other countries. The presence of mercury in the food supply has been reported widely in the popular press in Canada and the United States, and for many years guidelines tailored to specific lakes have been available to recreational fishermen in Canada and, more recently, in the United States.

Mercury in Fish of the James Bay Territory

Since the early 1980s, concentrations of mercury have been monitored in fish found in the natural lakes of the James Bay region as well as in the reservoirs of the La Grande complex and the rivers of the western part of the region. Three of the five Cree villages on the shores of James Bay and one of the four inland Cree villages are in close proximity to these waters (see map 3, xix). The monitoring has concentrated on four species of fish used by the Cree: the non-predatory longnose sucker (*Catostomus catostomus*) and lake whitefish (*Core gonus clupeaformis*), and the predatory Northern Pike (*Esox lucius*) and walleye (*Stizostedion vitreum*), which feed on smaller fish. For comparison purposes, and because it is known that mercury accumulates over time and with the age of fish, the samples look at fish of standard size. The edible flesh of each species is also sampled to provide a measure that would reasonably reflect human consumption.

Because of the elevation of mercury in fish in the reservoirs of the James Bay region and the attendant health concerns, the James Bay Mercury Agreement was signed in 1986 by the Cree Regional Authority, the Grand Council of the Cree, the nine bands of James Bay Cree, the Quebec government, Hydro-Québec, and the Société d'énergie de la Baie James. The ten-year Agreement provided for extensive monitoring of mercury in the environment and its impact on the Cree. Mercury in fish and wildlife, especially certain target species of fish in the reservoirs, diversions, rivers, and natural lakes of the region, is regularly monitored.

The elevation of mercury concentrations in fish harvested from the reservoirs and streams associated with the James Bay hydroelectric development have been well documented by the James Bay Mercury Committee (1992–93, 1993–94, 1994–95). The western sectors of the La Grande complex were flooded in 1979 and by 1981–82 there was unambiguous evidence that mercury levels had risen by a factor of three or four in lake whitefish and five to seven in northern pike. In adult fish, these concentrations have remained fairly constant since the 1980s, whereas in smaller fish of these species, the mercury concentrations have peaked and are slowly decreasing (James Bay Mercury Committee 1994–95). Models based on reservoirs of various ages in northern latitudes indicate that levels of mercury in the aquatic system, and especially in the fish, will continue to decline and reach background concentrations in several decades.

Representative mercury concentrations for standardized fish sizes are given in Table 3. Mercury is present even in fish from the natural lakes of the region. Although the levels are quite low, the average

Table 3   Representative concentrations of mercury in the flesh of important fish from the western regions of the James Bay Hydroelectric Development in 1992 (James Bay Mercury Committee 1992–93; 1993–94)

| | Concentrations in milligrams (mg) of mercury per kilogram (kg) of fish (parts per million) | | | |
| | Non-predatory species | | Predatory species | |
Location	Longnose Sucker (400 mm)	Lake Whitefish (400mm)	Northern Pike (700 mm)	Walleye (400 mm)
Natural lakes	0.2	0.2	0.6	0.7
Rivers				
Eastmain (residual)	0.2	0.1	0.8	0.8
La Grande (below La Grande 2)	1.0	1.2	3.3	—
Reservoirs				
La Grande 2	0.5	0.5	3.5	2.5
Opinaca	0.6	0.5	2.7	2.0
La Grande 3	0.7	0.4	3.4	—

northern pike and walleye from these lakes have mercury concentrations that slightly exceed Canadian governmental guidelines of 0.5 parts per million. While the data are not shown in Table 3, the same is true of the residual flow in the Eastmain river below the diversion. Fish from the La Grande River below the second La Grande dam and also from the La Grande estuary at James Bay have elevated mercury levels. Longnose sucker, lake whitefish, and northern pike found in these waters average about five to six times the mercury concentration of similar-sized fish in the natural lakes. In the major reservoirs, the two non-predatory species have mercury concentrations that are two to three times higher than background concentrations, putting them approximately at the Canadian governmental market standard of 0.5 parts per million. In these same reservoirs, mercury concentrations in predatory species exceed natural mercury concentrations by three to six times and recommended Canadian guidelines by the same amount. The general relationships shown in Table 3 seem to hold for the newer reservoirs of the James Bay development (James Bay Mercury Committee 1993–94, 1994–95). The following general pattern seems to be emerging: mercury concentrations are higher in fish caught in the reservoirs, regardless of species and feeding habits, than in fish of the same species caught in natural waters. The increase in mercury levels in non-predatory fish peaks within five years and then gradually declines. The increase in predatory fish species peaks at higher concentrations and appears to take ten years or so to reach the maximum. While mercury levels in the predatory species do decline, the phenomenon is less well documented because it has only recently begun to occur in the oldest of the James Bay reservoirs.

The James Bay Mercury Agreement supports research aimed at discovering and understanding the ultimate source of the mercury and the toxicity of mercury to wildlife species (such as mink, otter, and osprey). These measures include examining ways to prevent mercury from accumulating in reservoirs and becoming available to fish. Efforts have been made to improve methods of analysis and increase understanding of the processes of mercury methylation and biological fate.

As noted, mercury exposure among the Cree of the region is also monitored. Surveys of the fish consumption of the Cree were instituted, as well as a program to track body burdens of mercury in all segments of the Cree community, but especially in women of reproductive age and those who regularly consume large quantities of fish. The latter include individuals who maintain a traditional way of life such as hunters and trappers. Epidemiological research programs have been undertaken to determine whether the chronic low level of mercury ingestion has any adverse effects.

The agreement also provides for the evaluation of the sociocultural and economic impacts of increased awareness of elevated mercury exposure and of ways to mitigate the actual or potential impacts. Educational materials are distributed among the Cree regarding elevated concentrations of mercury in fish and related health concerns. In recent years public authorities in the United States and Canada at the state, provincial, and federal levels have posted additional advisories more stringent than those based on the usual ADI levels, cautioning recreational fisherman not to eat too much fish from certain waters because of the elevated levels of mercury or other pollutants in specified fish stocks. This same approach has been taken by the James Bay Mercury Committee in providing advisories to the Cree of the James Bay region. A video, posters, and numerous other educational materials have been developed. The committee has recommended dietary changes such as relying more on whitefish and other non-predatory species and less on predator fish such as pike, as well as increasing reliance on non-fish resources such as geese. These advisories have been widely advertised and distributed and they raise cautions about certain fishing areas and suggest amounts and types of fish that can be eaten by various subpopulations in the region.

Health Effects among the Cree

Before the reservoirs of the James Bay hydroelectric development were completed and filled, epidemiological studies were done on Cree adults (McKeown-Eyssen and Ruedy 1983) and young children (McKeown-Eyssen et al. 1983) from northern Quebec, including the James Bay region, to document the potential effects of methyl mercury intoxication. Based on medical examinations, adults were classified according to whether or not they had abnormal neurological traits. Mercury levels in blood and hair were determined independently. The study found a positive association between these mercury values and neurological abnormalities for adults of the inland village of Mistissini. Adults in Great Whale (Kuujjuarapik and Whapmasgoostui) showed a positive association between neurological abnormalities and increasing mercury levels in their tissues, but the association was not statistically significant. Whether these results are of biological significance is still debated by experts. The levels of mercury in hair samples ranged from about five to twenty-five parts per million, which is just at and above the level specified as safe by the World Health Organization but well within the safety factor of ten that had been applied to the lowest observed neurological effects. For example, the lowest exposure associated with neurological disease in Iraq had hair mercury concentrations of 120 to 600 parts per million. Thus, it appears unlikely that the observed neurological abnormalities can be attributed entirely, if at all, to mercury.

For children aged twelve to thirty months at the time of the neurological analyses, maternal methyl mercury levels in hair provided inconsistent results (McKeown-Eyssen et al. 1983). Maternal hair levels of methyl mercury were similar for mothers of boys and girls and averaged six parts per million in hair. Only six percent of the children were exposed in utero to levels greater than twenty parts per million. Of several neurological parameters, only one, abnormal muscle tone or reflexes, was significantly associated with maternal hair levels of methyl mercury for boys. For girls, there was no significant positive association between exposure and neurologic abnormalities and poor coordination was negatively associated with exposure. There was no consistent dose-response relationship. Because of the inconsistency between the findings and the dose-response relationship, the authors were very careful to indicate the tentative nature of the studies.

Annual reports concerning the Cree and their exposure to mercury have been provided by the James Bay Mercury Committee, as stipulated by the James Bay Mercury Agreement. Since 1982 monitoring of the mercury status of the Cree has come under the auspices of the

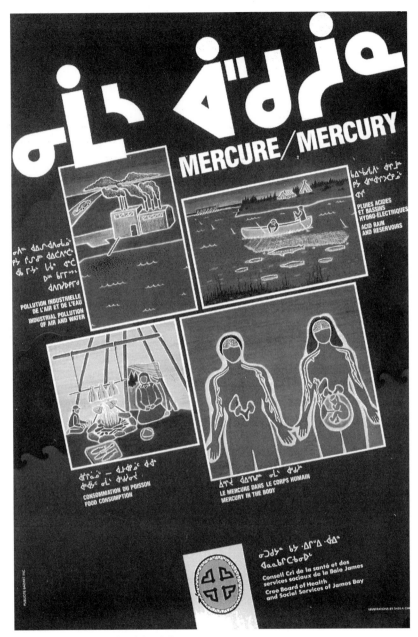

Figure 1   Mercury health advisory

Cree Board of Health and Social Services of James Bay. From their inception, these mercury monitoring programs have focused upon target populations – those with the greatest exposures to mercury (mainly the elderly, trappers, and others living in the bush for considerable periods) and women of childbearing age (fifteen to thirty-nine years), who are of concern because of the association of high body burdens of mercury with birth defects. Mercury content in hair is the primary measure that has been employed. The medical authorities of the Cree Board of Health have decided that hair mercury values above thirty parts per million are of potential concern for all except women of childbearing age, for whom the threshhold of concern is set at fifteen parts per million. It is safety that is being monitored, and these mercury values fall below concentrations that are known to be toxic or associated with birth defects.

The mercury monitoring and education programs seem to have been successful in reducing the concentrations of mercury in the Cree to levels that are below the values that cause concern. In the general population, the body burden of mercury as reflected by hair values of mercury has decreased in the last decade; in fact, people under twenty years of age rarely exceed the limit of mercury detection (2.5 parts per million) in hair. This clearly indicates that the ingestion of fish in the young is small to non-existent. Likewise, mercury burdens have decreased substantially in mothers and their newborns (James Bay Mercury Committee 1993–94). There remain some concerns about the generally higher concentrations of mercury in women whose spouses are trappers. Cree living in the four inland villages have levels of mercury similar to those in age-matched groups from the coastal villages, with the apparent exception of inland residents forty years of age and older, who consistently had somewhat higher hair mercury values (ibid. 1993–94, 1994–95); apparently, more fish is being ingested in these areas, and the situation is being closely watched by the mercury committee.

Additional data will be gathered from monitoring as time passes, including further information on the effects of mercury exposure in children between four and fifteen years. To date most of the monitoring has been undertaken in the late summer, but it is now known that major exposures to mercury also occur in the late fall; hence the monitoring schedule is being reconsidered. This refinement of the monitoring program is unlikely to lead to major new discoveries. Nor is the increased surveillance of children likely to reveal anything dramatic since it is quite clear from the monitoring program that the young simply do not have a high mercury burden. For example, mercury values in hair are at undetectable levels in those younger that twenty years of age. In older people the mercury values increase and are highest in the eldest members of the population. It seems reasonable to predict that as the population ages, mercury in hair will decrease, not because the level of mercury in fish is slowly declining but because fish is no longer accepted as a food item by the newer generations.

Unfortunately, the quality of the James Bay Mercury Committee reports is difficult to judge since they contain only summaries of the database and lack sufficient detail. About seventy-five percent of those in the various target populations agreed to be studied, but without specifics it is difficult to determine whether this sample is adequate or not. The main concern with the monitoring-program data is the reporting of the results. Most of the data can be found in various reports of limited distribution, not in the broader peer-reviewed scientific literature. While widespread publication of its data is not the main focus of a monitoring program, knowledge and acceptance of its findings among the worldwide scientific community could only strengthen the credibility of the programs. Moreover, the data could be valuable to others studying and monitoring human health.

The second main concern with the data is that they do not address the relationship between types and amounts of fish eaten and the subsequent mercury values. Since the observed decrease in mercury may in part be attributable to the fact that some individuals have simply stopped eating fish, data of this sort are needed, along with some measures or estimates of the value (cultural and economic) of fish to the community. For example, having a mercury monitoring value indicating that there is no mercury exposure may actually be undesirable from an overall health perspective; it would be better to accept a certain amount of mercury since it would indicate some ingestion of fish.

Social Impact on the Cree

The James Bay Mercury Committee, through the organizations and villages with which it has worked, has developed various measures to mitigate the increased mercury content of fish. In some ways the education programs may have been all too successful in changing the dietary practices of the Cree. Figure 1 gives an example of an educational poster. In this area, too, the data that exist are sketchy and have not been aired in peer-reviewed scientific literature. It is important that further studies be done.

Mitigative measures have focused upon encouraging substitutes for the foods that contain the highest amounts of mercury. Monitoring fish catches and controlling the mix of predator and non-predatory fish species, for example, have led to the establishment of community-based programs to exploit lake whitefish stocks more than the predator species such as pike and lake trout. Similarly, the development of fisheries in coastal James Bay has been subsidized. The harvesting of geese, while contain little mercury, has been enhanced through habitat modification such as the building of ponds that geese might use and the clearing of trees around certain existing ponds. There are no available data on the success of these measures, nor even simple dietary surveys to determine whether geese are being eaten more frequently. Aerial surveys have been made to locate beaver lodges along traplines. Presumably the surveys would encourage further trapping, but even if an increase does occur it is not clear what impact it would have on the ingestion of mercury by the Cree.

There seems to be limited access to a variety of foods in the James Bay region. The high-profit foods, that is, the snack food items such as soft drinks, candy, and chips, are often in greatest abundance. Many of these ready-to-eat foods are high in salt and fat and low in vitamins, minerals, and protein. High consumption of these foods, coupled with the generally sedentary lifestyle in the villages, creates an unhealthy environment. It would be interesting to compare the dietary quality and nutritional status of Cree who have virtually no mercury in their hair samples, and those who show the presence of mercury, but at levels that are below the threshold of concern. Such data would help to determine the impact of not eating fish.

Other public health alarms surrounding food items have arisen in recent years. Non-nutritive sweeteners such as saccharin were linked to bladder cancer; apples and apple products with traces of Alar™ raised a risk of cancer in children; coffee ingestion was linked to cancer of the pancreas. Each of these scares subsided in time. But the issue of mercury and the Cree is different in that there is a ready and easy way to monitor both exposure to mercury and the results of the campaign to educate and inform those at risk from mercury ingestion. This campaign, with graphics such as the poster in figure 1, has undoubtedly affected the patterns of fish consumption. Indeed, it appears that the young may be rejecting fish as a suitable food and replacing it with a less-healthy diet. It is important to document this possibility and its impact upon the Cree.

Conclusion

All food supplies contain undesirable chemicals such as mercury. Because mercury is naturally present in the environment, it has undoubtedly been present in (i.e., has contaminated) fish throughout the world for eons. It was present in the fish of the James Bay region when the ancestors of the present-day Cree first migrated to the area. What is different now is a general and widespread elevation in the concentration of mercury in fish due to global and regional industrial activity, and localized elevations from the formation of reservoirs.

Measuring the increased mercury concentration in fish is relatively easy and is being done. Measuring the dietary burden on the Cree is more difficult, measuring the resultant health effects more difficult still. In the James Bay region it appears that the increased mercury concentrations in fish are such that some people, whether because of diet, age, or other special circumstances, may be seriously affected if they do not change their diet. It is possible that this was true, though for a much smaller number of people, even before the reservoirs were flooded since "natural" mercury concentrations in some areas may already have been high. It also seems likely that relatively modest changes in dietary behaviour, such as those suggested by the James Bay Mercury Committee, would greatly reduce the number of people who are seriously affected. Simply limiting the consumption of predator fish, especially by young children and women of reproductive age, might well reduce the risk to an acceptable level.

Human response to risk, even where the risk is fairly well understood, is frequenty determined by factors beyond statistical likelihood and the severity of the risk itself. The most extreme responses are triggered by risks that are new, invisible, and poorly understood. Among the Cree there is now a general perception that the fish supply has been in some sense poisoned. The image of an invisible contaminant in a staple food, that is also a cultural anchor has caused an extreme and entirely understandable reaction of fear that is nonetheless out of proportion to the level of the risk itself. In itself an extreme reaction is not harmful unless it leads to alternate behaviour that is harmful.

There are many risks associated with a major shift away from fish as a dietary staple. Fish is an extremely nutritious food, high in protein, low in fat. If fish consumption were reduced and other high-quality foods filled the void, then the recommendations for reduced or restricted fish consumption could be beneficial. However, if a fear develops that all fish in the James Bay region are "poisoned" and poisonous regardless of the amount of mercury present in their flesh, then a widespread replacement of fish will likely result in poorer nutrition. Similar nutritional concerns have been raised with respect to the restrictive fish consumption advisories recommended by the United States Environmental Protection Agency (Egeland and Middough 1997). Because of mercury in fish taken by recreational fishermen, the guidelines have the effect of limiting use of an exceedingly nutritious food by substance fishermen.

Fear of fish also carries an economic and cultural price. Foods imported from outside the region are expensive to purchase and the availability of a wide selection of high-quality nutritious foods in north-ern stores cannot be assured. To the Cree, fish are free (except for the initial outlays for equipment and transportation to fishing areas) and may be freely harvested. Culturally, fish has been an item of consider-able significance to the Cree, both as a food and as an item of com-merce. Fishing reinforces traditional activities; it provides a connec-tion to the past and a linkage to the land. Given the cumulative cultural impact of the James Bay project on the Cree, preservation of the cultural fishing tradition may deserve high priority indeed.

Without a doubt, fish is an exceedingly valuable food for the Cree. Its excessive devaluation from an unreasonable fear of mercury poi-soning is of serious concern. The dose of mercury (which depends on the type and quantity of fish consumed) must be balanced with the nutritional alternatives and risks attendant on the consumption of other foods. The appropriate approach would be to acknowledge the issues and uncertainties surrounding the levels of mercury contami-nation in fish and its consumption. Thereafter, with education, the population can make rational decisions regarding the use of fish.

Toxicologists and nutritionists would generally agree with the physi-cian and alchemist Paracelsus (1493–1541) that "all substances are poisons; there is none which is not a poison. The right dose differen-tiates a poison from a remedy." In the context of James Bay, that advice might be applied not only to the problem of enhanced mercury con-centrations in the diet of the Cree but also to the various perceptions of, and reactions to, the problem.

Acknowledgments

I thank Jean Hennessey of the Institute on Canada and the United States, Dickey Endowment, Dartmouth College, for her encourage-ment, interest, and numerous contacts with all parties. The research and travel involved in writing this chapter was supported by a Quebec Research Grant through the Quebec Delegation in New England.

REFERENCES

Bodaly, R.A., R.E. Hecky, and R.J.P. Fudge. 1984. "Increases in Fish Mercury Levels in Lakes Flooded by the Churchill River Diversion, Northern Mani-toba." *Canadian Journal of Fisheries and Aquatic Sciences* 41, no.4:682–91.

Clarkson, T.W. 1991. "Inorganic and Organometal Pesticides." In *Handbook of Pesticide Toxicology*, edited by W.J. Hayes and E.R. Laws. New York: Academic Press.

Egeland, G.M., and J.P. Middough. 1997. "Balancing Fish Consumption Benefits with Mercury Exposure." *Science* 278:1,904–5.

Gorrie, P. 1990. "The James Bay Power Project." *Canadian Geographic* 110, no. 1 (February/March):21–31.

Hamilton, A., and H.L. Hardy. 1974. *Industrial Toxicology*. Acton, MA: Publishing Sciences Group.

Hunter, D., R.R. Bomford, and D.R. Russell. 1940. "Poisoning by Methylmercury Compounds." *Quarterly Journal of Medicine* 9:193–213.

James Bay Mercury Committee. 1992–93. *Report of Activities*. Sixth Annual Report. Montreal: James Bay Mercury Committee.

– 1993–94. *Report of Activities*. Seventh Annual Report. Montreal: James Bay Mercury Committee.

– 1994–95. *Report of Activities*. Eighth Annual Report. Montreal: James Bay Mercury Committee.

Marsh, D.O. 1987. "Dose-Response Relationships in Humans: Methyl Mercury Epidemics in Japan and Iraq." In *The Toxicity of Methyl Mercury*, edited by C.U. Eccles and Z. Annau. Baltimore: Johns Hopkins University Press.

McKeown-Eyssen, G.E., J. Ruedy, and A. Neims. 1983. "Methyl Mercury Exposure in Northern Quebec II. Neurologic Findings in Children. *Journal of Epidemiology* 118:470–9.

McKeown-Eyssen, G.E., and J.M. Ruedy. 1983. "Methyl Mercury Exposure in Northern Quebec I. Neurologic Findings in Adults. *Journal of Epidemiology* 118:461–9.

Mitchell, J.G. 1993. James Bay: "Where Two Worlds Collide." *National Geographic* 184, no.5 (November):66–75.

National Research Council Canada. 1979. "Effects of Mercury in the Canadian Environment." In *NRCC Associate Committee on Scientific Criteria for Environmental Quality*. Ottawa: National Research Council Canada.

Ramade, F. 1987. *Ecotoxicology*. New York: John Willey and Sons.

Takeuchi, T., F.M. D'Itri, P.V. Fischer, C.S. Annett, and M. Okabe. 1977. "The Outbreak of Minamata Disease (Methyl Mercury Poisoning) in Cats on Northwestern Ontario Reserves." *Environmental Research* 13, no.2:215–28.

Tollefson, L., and F. Cordle. 1986. "Methylmercury in Fish: A Review of Residue Levels, Fish Consumption and Regulatory Action in the United States." *Environmental Health Perspectives* 68:203–8.

Verhovek, S.H. 1992. "Power Struggle." *New York Times Magazine* (12 January):16–27.

Weiss, M.K., M. Bertine, M. Koide, and E.D. Goldberg. 1975. "Chemical Composition of Greenland Glacier." *Geochimica et Cosmochimica Acta* 39:1–10.

Wheatley, B., A. Barbeau, T.W. Clarkson, and L.W. Lapham. 1979. "Methylmercury Poisoning in Canadian Indians – The Elusive Diagnosis. *Journal Canadien des Sciences Neurologiques* 6, no.4: 417–22

World Health Organization. 1990. *Environmental Health Criteria 101: Methylmercury*. Geneva: World Health Organization.

Xun, L., N.E.R. Campbell, and J.W.M. Rudd. 1987. "Measurements of Specific Rates of Net Methyl Mercury Production in the Water Column and Surface Sediments of Acidified and Circumneutral Lakes. *Canadian Journal of Fisheries and Aquatic Sciences* 44, no.4: 750–7 reflecting a combination of slack energy demand, significant political protest, and the coming to power of the Parti Québécois in Quebec. (For a better visual perspective, consult the maps on pages xvii–xix.)

Adjacent to Whapmagoostui is an Inuit village, Kuujjuarapik, with nearly five hundred residents. It forms the boundary between the northern limits of Cree settlement and the southern edge of Inuit territory. To the north are two other Inuit villages, Inukjuak and Umiyjaq. Across the Passage de Manitounuk on the Belcher Islands is the Inuit community of Sanikiluaq, which was added to an extended environmental assessment area only after a political struggle. These are the primary Inuit communities in the impact area, although an additional six thousand Inuit live along Hudson Bay, Hudson Strait, and Ungava Bay. While this study primarily addresses the Cree, some of the issues and framework for analysis have relevance for the Inuit as well.

Traditionally the Cree have hunted, fished, and trapped throughout the James Bay region, dispersing into the bush in small hunting groups in the winter months and coming together in larger band gatherings during the short summer season. Over the past two or three generations the summer settlements have become increasingly permanent, with contemporary two- and three-bedroom housing and an infrastructure of schools, health care centres, water and sewage systems, police, cable television, a float plane transport service, recreational facilities, and band administrative offices.

We are at a point in history at which the accumulated injuries against indigenous peoples are increasingly acknowledged. The issue of the dams at James Bay is unavoidably a part of that larger reconsideration. This is not a simple case of eminent domain, in which some people are expropriated with fair compensation so that a larger number may benefit. Canadian courts have taken that position, applying the legal doctrine of "balance of convenience" in the James Bay context to justify the losses of a few for the benefit of the many. The formal theory of economic welfare begins with the proposition that if some persons gain and others lose from an economic action, then no unambiguous, *prima facie* claim of overall societal gain can be made. That position is amended to say that if the losers accept compensation from the gainers in an amount both parties find acceptable, then total social welfare is unambiguously higher. The Quebec government and Hydro-Québec interpret the 1975 James Bay agreement with the Cree within this framework. Not only was substantial compensation paid to the Cree but a three-tiered system of land use rights was also established. In rejecting this line of thought the Cree contend that the James Bay agreement never presented the option of refusing hydroelectric development outright; the agreement represents no more

5 The Cree People of James Bay: Assessing the Social Impact of Hydroelectric Dams and Reservoirs

STANLEY WARNER

The Setting

In 1975 the Cree, Inuit, Province of Quebec, Canadian federal government, Hydro-Québec, and two development corporations signed the James Bay and Northern Quebec Agreement. In the eyes of Hydro-Québec the Cree have been paid a total of over $450 million in return for the right to develop all three phases of the James Bay hydroelectric system. From the perspective of the Cree the agreement was signed under duress: construction of access roads had begun in 1971, work on the first dam had begun in 1973, and an injunction briefly halting construction had been overturned on appeal.

Approximately twelve thousand Cree live in nine communities distributed over this vast area. The villages range in size from four hundred to nearly three thousand. Five villages are on the eastern shore of James Bay at the mouths of rivers that are being dammed or diverted; the others are inland and to the south in areas that are thus far less directly affected by reservoir construction, although roads and powerlines cross this territory. The community of Chisasibi, formerly on an estuary island at the mouth of the La Grande River, was relocated in 1980 to the mainland as a precaution against possible erosion by the doubling of the river's flow. The village of Eastmain is located at the mouth of a river whose flow has been reduced by ninety percent due to diversion into the La Grande system. The northernmost Cree village, Whapmagoostui, lies at the mouth of the Grande Baleine River. Plans to dam the river were indefinitely postponed in 1994, than a "second best" response to unwanted development and intrusion. The recent 1993 negotiation of an additional $50 million for negative social impacts on the villages of Chisasibi and Wemindji is also interpreted by the Cree as recognition of and payment for past damages from construction of the La Grande complex and not as a signal that new development is acceptable (Larry House, telephone communication, 12 April 1993, and conversation with Violet Panochos, chief of Chisasibi, at Chisasibi, 20 July 1993).

This legal political framework wherein compensation is paid for the taking of lands as part of a social balancing of convenience is also rejected by a growing world movement that seeks to redefine the rights of indigenous peoples. (See, for example, the review of World Bank policy in Morse and Berger 1992, as it relates to the Sardar Sarovar dam project in the Narmada region of India.) Embedded in that reconsideration are matters of racism, economic exploitation, and cultural assault. Canadians claim a piece of the high ground in this arena, noting that their history of formal recognition of aboriginal rights and their social policy actions rank ahead of the records of most nations; but that does not remove the issue from the table.

Theory and the Anthropologists

A number of anthropologists, notably Richard Salisbury, Harvey Feit, Adrian Tanner, and Colin Scott, have specialized in understanding the Cree (Feit 1979, 1980, 1982, 1989; Salisbury 1986; Scott 1983, 1984, 1988, 1989; Tanner 1979, 1983). Their work offers varied windows on the evolving social order, livelihood, and culture of the Cree and seeks to define the patterns of interaction with so-called encapsulating societies. The effort by anthropologists and others to devise a theoretical framework for approaching the present circumstance of aboriginal or "first" peoples is itself in ongoing revision. One common thread is a general rejection of gradualist assimilation models, in which an "original" culture becomes incrementally Westernized (Bender and Morris 1991; Feit 1982). Instead, the initial (and ongoing) contact of native cultures with Euro-American or metropolitan societies is seen to produce distinctively new modes of production and social life – new formations that are not simply halfway points interpolated between the old and new.

In *Bringing Home Animals*, Adrian Tanner (1979), addresses the period of more than three hundred years in which fur trading with the French and English was established as a primary feature of Cree life. He challenges the gradual-acculturation model framed as a general theory by Murphy and Steward (1956) and developed specifically for northern aboriginal peoples by Leacock (1954) and Pothier (1967).

A "transformational" model of social change can be used to explain what happens following contact between an aboriginal population and a more powerful group of newcomers, and such a model is to be contrasted to a conventional "acculturation" approach to the same situation ... The transformational approach analyses the post-contact stage or stages as distinct from either that of the aboriginal society or of the dominant group. The post-contact society has a mode of production distinct from the aboriginal society due to technological innovations and trade acquisitions, but it is a mode of production that is also distinct from that of the dominant society. (Tanner 1979, 66)

Tanner finds parallels in this transformational stage with the "putting out" system developed in precapitalist Europe. Supplies for Cree hunting groups are advanced by the producer, in this case the Hudson's Bay Company, production occurs within the community, and the final product (furs) is purchased by the same enterprise. Tanner rejects Leacock's portrait of a society in conflict between trapping for trade and subsistence hunting (Leacock 1954), in which an individualized, more capitalist mode of production is displacing communal production, leading to a larger breakdown of Indian culture. He suggests "we stop looking at the Mistassini [a Cree community] as a group in conflict between traditional and modern elements, or between the contrary demands of hunting and trapping, and try to understand it as a social form in its own right" (Tanner 1979, 10).

Colin Scott carries this approach forward, examining the more recent period in which the commercial trading of furs has ceased to be a significant source of income for the Cree. The year 1983 was a benchmark; it was then that the European Economic Community, in response to protests by environmental and animal rights organizations, banned all imports of furs, triggering a precipitous fall in prices. It was the final chapter in a decline that had been underway since the 1960s. Earnings from the commercial trapping of furs dropped to two percent of all income (Salisbury 1986) and has not recovered since. In his article "Between 'Original Affluence' and Consumer Affluence: Domestic Production and Guaranteed Income for James Bay Cree Hunters" (1984), Colin Scott describes a dual economy comprising both year-round village residents, whose income is a combination of wages and government transfer payments (social security, unemployment compensation, etc.), and those who spend the long winter period dispersed in the bush, engaged in traditional subsistence activities.

This dual structure is supported by the income security program, a truly novel innovation established as one of the cornerstones of the 1975 settlement. Coupled with specific guarantees of territorial hunting rights, the Program sustains the practice of hunting, fishing, and trapping for personal consumption at a time when the loss of fur income would otherwise have brought a significant decline in these traditional activities. Each individual who spends more than 120 days a year in the bush is compensated by a formula that includes a per diem amount. Husbands, and wives, all receive separate payments. A typical husband and wife will receive $10,000 or more. These funds are joined with summer wage earnings and other transfer payments. Richard Salisbury (1986) estimated the imputed value of subsistence food production by the James Bay Cree to be twenty percent of community income from all sources. It should be noted that even those Cree who are not regular bush hunters still participate in the big spring goose hunt and other part-time hunting and fishing.

More important than the imputed value of subsistence food production are the social relations that revolve around an egalitarian ethic of sharing. Hunting surpluses are shared widely with non-hunters, who reciprocate at holidays with substantial gifts of cash or valued goods. Scott argues that within the dual character of the Cree economy,

a guaranteed annual income for Cree subsistence producers has led to more hunting households being active in the bush for longer periods of the year, and a reassertion of traditional production units ... The Income Security Program was introduced at a time when the growth in incomes of those permanently employed could have led to a break-down in egalitarian exchange and to permanent stratification ... Continued subsistence production is an anchor for the maintenance of traditional social forms, while involvement with diverse forms of urban society is expanding. (Scott 1984)

Thus, for Tanner and Scott (and Feit as well), Cree society has evolved in transformational stages that have incorporated new sources of income and livelihood while preserving a matrix of cultural values and practices from earlier times. When the Cree, in their opposition to Hydro-Québec, speak of protecting their traditional way of life, the reference is more likely to be to securing its more recent transformational state than to recovering an earlier form.

Some Euro-Americans who have organized in support of the Cree carry a stylized vision of Cree life that sets up an oppositional relationship between an original balanced-with-nature culture and the bulldozer forces of an intruding Western culture. In a form that echoes the "noble savage" image, the Cree are sometimes represented as a timeless people unable to cope with the forces of modernity. David Suzuki, for example, producer and host of films on nature and social issues, offers a portrait of the Cree that is largely devoid of the elements of modern technology and social life: we are shown an old man in traditional handmade clothing kneeling in the snow setting a rabbit snare, rather than someone travelling to the hunt by snowmobile or in a seventy-horsepower outboard canoe, wearing a polyester hunting jacket and carrying a modern automatic rifle or shotgun. An interview with the Cree about the hydro project is conducted by Suzuki in a traditional family bushcamp structure constructed of bent wood wrapped in an outer skin, with spruce bows for sitting arranged around an open fire. Less emphasized are the interiors of typical Cree homes within a community, with children watching television or playing Nintendo, or the formal administrative offices within which the Cree coordinate a range of public programs. If one walked through a Cree community such as Chisasibi or Mistissini with a video camera panning right and left (with an appropriate sense of voyeur's guilt for doing so), the resulting footage would be more complex and ambiguous than the selected images of traditional life that Suzuki assembles; for in reality traditional and modern have been joined in ways that continue to carry a strong signature of Cree identity and social structure. If well-intentioned individuals and organizations come to the support of the Cree with a constructed vision of a primitive, simple people, they do disservice to the intellect of first peoples and risk a subtle form of racism. Better than framing an argument around cultural vulnerability or the protection of an "original" culture is an approach that addresses the unequal power relations that inhibit the search for self-determination and local control.

In the analysis of aboriginal cultures, anthropologists have also struggled to develop a more endogenous account of social change. Barbara Bender and Brian Morris, in their review of "Twenty Years of History, Evolution, and Social Change in Gatherer-Hunter Studies" (Ingold 1991), note the tendency, right through the 1970s, to view precontact societies as timeless and pristine and to locate all postcontact change as externally imposed. "The Cree, the San, the Inuit, and the Mbuti," they reply, "have not been preserved in amber; they are no more 'pristine' than any other gatherer-hunter society; indeed, their flexibility, tendency to fission and low-profile tactics may in part be a form of resistance to domination ... There can be no justification for assuming that one set of attributes is 'timeless', another the result of exogenous forces" (Bender and Morris 1991, 7).

Also in question is a companion impulse to specify a universal quintessential "foraging mode," within which property is communal, the sexual division of labour is rational, leadership is horizontal, gender inequality is minimal, kinship networks predominate over nuclear families, and an environmental ethic manages the natural world. "Original" societies then become a foil for the critique of overdeveloped postindustrial modernism. In the anthropological meetings of the 1980s researchers began to give hunter-gatherers their own history, to understand them as differently constituted, and to question models that portrayed them as responding passively to external forces. Bender and Morris conclude: "Above all, the message of this volume must be that gatherer-hunters have their own history. An understanding of the processes of encapsulation has to work in tandem with an understanding that gatherer-hunter variability, past or present, has an internal dynamic. Change in gatherer-hunter societies does not wait upon the arrival of land-hungry farmers, nor upon capitalist penetration" (ibid., 13–14).

Harvey Feit, who has written widely on the Cree and who served as an anthropologist advisor to the Cree at the negotiation of the James Bay agreement, brings this issue more directly into focus. Hunting is a central activity that gives shape to many of the social and cultural relations of Cree life. Feit details the degree to which, prior to 1975, the Cree had been losing control of hunting rights and the management of wildlife resources (Feit 1982). In the 1930s, spurred by high fur prices, white hunters and trappers encroached upon Cree hunting grounds. The Cree responded by pre-emptively increasing their own harvest, driving the beaver to near extinction. The Canadian and Quebec governments intervened in ways that brought a "unilateral redefinition of native hunting rights," denied "the existence of an aboriginal right to hunt," and fostered an increased dependence on the governmental welfare system (ibid., 397–8). In 1975 the James Bay agreement reversed the pattern. The Cree came to the negotiations seeking solutions to five related issues: "definition and recognition of aboriginal hunting rights; effective involvement of the Cree in management of wildlife and the environment; regulation of the allocation of wildlife between indigenous hunters and sportsmen including priority allocation to the former; regulation of the environmental impacts of developmental activity; and reduction of the dependence of the hunters on world-market conditions and on government policies and programs" (ibid., 396).

The resulting agreement strengthened subsistence production, reduced welfare dependence, and "clearly demonstrated how a hunting people can undertake to define their own futures" (ibid., 402). "The acculturation model, and indeed many other anthropological theories of change, involves the assumption that micro-level changes originate primarily externally, and more particularly that local-level responses are simply reactive, the local population having neither the power nor the means to generate unique or effective responses ... It is clear ... that hunting societies have not been simply passive in the face of external changes, and that many have sought to set and meet their own objectives" (ibid., 403).

Feit closes his analysis with a series of prescriptive conclusions. We need to "[recognize] that local values and knowledge play a central part in the process; revise our theories of change so as to better account for local-level power; and [construct] a critique of developed nation-states that does not fall into the structural trap of seeing them as all-powerful and hyper-integrated" (ibid., 404–5).

With the passage of time Harvey Feit became less persuaded of the gains from the structural features of the 1975 agreement and more concerned with the inadequacies of enforcement: "Large-scale industrial development projects are continuing on Cree lands ... The failure to adequately regulate development is a major future threat to the revitalized hunting sector. These threats demonstrate ... that the agreement process was unable to resolve fundamental conflicts between the interests of Cree and those of wider economic and political institutions of the capitalist economy or the liberal democratic state" (Feit 1989).

The discussion to this point has centred upon the larger theoretical issues of transformational versus acculturational models and externally driven versus endogenous sources of change. It is time to move to a more specific, concrete level of impact analysis.

Social Impact

At various times the Cree (and Inuit) have spoken of their historic rights and claims to the land. Yet strategically they have advanced their case on environmental grounds – or, more precisely, on the social consequences of environmental changes. The transformation of the natural environment (from the effects of reservoirs, roads, and powerlines) is identified as the central contemporary force eroding and dismantling their traditional way of life (Coon-Come 1991; Masty 1991).

This distinction between arguments grounded in principles of territorial self-determination and arguments grounded in the social impact of environmental change is important. The analysis that follows begins with the environmentally grounded approach. It inquires into the extent to which hydroelectric development, by transforming the physical landscape, thereby transforms the quality of Cree life. Such an inquiry does not encompass or resolve the constitutional debate on autonomy and self-determination for indigenous peoples. Who controls these lands and for whose benefit is a question to be revisited.

The structure of Cree communities, the sources of Cree livelihood, and the character of Cree culture have been changing significantly – especially in recent decades – and will continue to evolve even in the absence of further hydroelectric development. What is the differential impact of the dams? How are both the quality of change and the rate of change to be distinguished?

Demographic factors are significant. Various records indicate that between the 1870s and the 1920s the population of Eastern Cree in the James Bay region was relatively stable, in the range of 1,400 to 1,800 persons (Coon-Come 1991; Masty 1991). This number, or fewer, may be indicative of the long-run equilibrium level over prior centuries. But by the mid-twentieth century a major expansion of population was underway. Declining infant mortality and increased longevity, coupled with a relatively high birthrate, brought a more than doubling of population to over 5,000 by 1970. Since then the number of Cree has more than doubled again, to nearly 12,000. Approximately fifty percent of the Cree are now under the age of twenty; yet another doubling is likely within the next thirty years. Such rapid growth is not out of line with world patterns of population expansion, but it raises specific issues about traditional hunting and trapping in relation to carrying capacity and as well, may contribute to community stress. The Cree are concerned about the options and difficulties facing this large cohort of young people, including problems of teenage pregnancy, drug abuse, loss of traditional values, and uncertainty about future livelihood choices.

As noted, hunting and fishing for food are still a significant dimension of Cree life. Anthropologist Richard Salisbury estimated that the imputed value of game and fish harvests represented fifty-seven percent of all income in 1971 and twenty percent in 1981. The decline does not reflect diminished hunting so much as a sharp increase in income from other sources (Salisbury 1986). A study of eight villages over the period 1972-79 reports on the variety of animals that were hunted. The average annual Cree harvest included 63,000 Canada geese, 48,000 ducks, 21,000 snow geese, 3,600 loons, 17,000 beaver (eaten as well as skinned), 42,000 hare, 900 moose, 800 caribou, and 200 black bears (Beaulieu 1984). (For comparison, Hydro-Québec's wildlife biologists estimate that the La Grande reservoirs displaced or eliminated 7,000 to 9,000 nesting pairs of migratory waterfowl.) Fishing, an important summer activity, contributes about twenty-five percent of animal protein to the Cree diet. Whitefish, a bottom-feeder that consumes small invertebrates and has a lower propensity to accumulate mercury, is preferred, although a range of predator fish, including the salmonids and pike (in which higher levels of mercury are recorded) are also eaten. (For a detailed analysis of the elevation of mercury levels in hydroelectric reservoirs, see chapter 4.)

The combination of rapid population growth and the importance of hunting and fishing is producing a demographic dilemma. The traditional Cree way of life has historically meant a balance between the number of Cree hunters and the carrying capacity (i.e., reproductive capacity) of the land. Elaborate social customs determine which commensal groups control the hunting rights to particular territories and how permission is granted for seasonal or occasional use to those who do not normally hunt in a given area (Feit 1982; Tanner 1979). The Cree underscore that they practise responsible husbandry and stewardship of the land. However, a growing population means a smaller percentage of young Cree can expect to enter traditional hunting and trapping areas to take beaver, fox, lynx, martin, moose, bear, and waterfowl. (Moreover, a growing number of non-Cree hunters are pressuring the same resources.) Thus, a rising percentage of young Cree must find a different livelihood. With half the population under age twenty, the issue must be addressed in this generation.

One of the objectives of the 1975 James Bay agreement was to establish a capital endowment fund. Launched with an initial $150 million for basic compensation and $75 million as compensation for future development (Beauchemin 1992), the fund is an investment portfolio whose earnings are intended to finance new patterns of Cree economic development. One venture, Air Creebec, provides regional scheduled airline service, but over the years it has operated at a loss and imposed a significant net drain on capital funds. The construction of new housing provides summer work that meshes nicely with people's return from the winter hunt. Although spurred by population growth, housing construction will not provide the same proportion of future jobs it has generated in the last decade. The marketing of native arts and crafts has been supported, but it is unlikely to employ large numbers. Professional guide service for tourist hunters and fishermen is an option only lightly explored, perhaps because it carries a bittersweet or repugnant sense of escorting hunters for sport on to lands that are historically identified with the life-sustaining activities of particular families. A processing plant has been established for the commercial marketing of fish. Hydro-Québec established training programs for the Cree for hydroelectricity jobs; the number of participants was small. An important source of wage employment is associated with community infrastructure, including the schools, the clinics, the police, and band administrative offices. But overall, as the number of working-age adults doubles and the carrying capacity of the land for subsistence activities remains finite, alternative opportunities for work seem to be insufficient to meet the need. No coordinated plan or collective vision appears to be emerging.

The pattern of subsistence production identified by Scott and Feit as the successor transformational stage to the fur-trading era that preceded it will come into question. At present, twenty-five percent of the Cree are now engaged in winter hunting and trapping, dispersed in the bush and assisted by the income security program (Cree Hunters and Trappers Income Security Board 1993). While a minority, their activities nevertheless constitute a form of social glue that holds together the traditional culture of the community through a variety of linkages (Feit 1982; Scott 1988; Tanner 1979). How much farther can the number of full time hunting families fall before this kind of glue (and transformational model) gives way to a different construct? Thus far the anthropological analysis has featured the twenty-five percent who can be described as gatherer-hunters but has said little about an independent social role for the seventy-five percent who remain, and whose numbers will grow.

In addition to the dynamic set in motion by demographic forces, a spectrum of factors are changing the character of Cree culture on an ongoing basis. In the past twenty years technological changes have reshaped both hunting and daily community life. Motorized transport – whether by air or water, on snow or newly constructed roads – is the preferred mode of travel. Satellite dishes bring multiple television channels. Contemporary housing (often with electric heat) is producing its own architectural history, as styles and tastes change. Schools and health services incorporate developments from the wider society. Thus, when speaking of the entire culture, economy, and social organization of a people, it becomes risky and difficult to assign social changes to particular causes or events such as hydroelectric development. The Cree have tended to minimize social dislocation brought on by the larger sweep of events and have instead emphasized reservoir building as the most culturally disruptive force in their lives. And, as noted earlier, supporters of the Cree often project the image of an original and pristine culture about to be decimated.

One way of resolving this incongruity is to remember that the Cree have not simply had a broad spectrum of technological and social change thrust upon them but have exercised a significant measure of agency or choice in what is to be adopted, the terms of its introduction, and the shape of its use. By contrast, the undertaking to dam the major rivers of James Bay and create a comprehensive system of reservoirs, roads, and powerlines is invasive and intrusive.

It should be noted, however, that while attention is often focused on the degree to which the traditional ways of the Cree are overwhelmed by contemporary forces, the record is not entirely negative. In several respects Cree life has gained in integrity and vitality. Education is now centred in locally controlled schools, in contrast to earlier compulsory education practices whereby children were shipped to remote boarding-schools. The Cree language is supported rather than suppressed within the educational system. English as well as Cree is spoken in the schools, and English continues to be the principal second language for the Quebec Cree. This has political ramifications for Cree relationships with federal Canada and francophone Quebec. The infrastructure of local services and governance is principally controlled by the Cree. The income security program has encouraged continued hunting and trapping in the bush in the winter months.

On a larger scale – funded by financial settlements and impelled by the struggle with Hydro-Québec – the Cree have developed an overarching political structure. The Grand Council of the Cree, with offices in Ottawa and Montreal, coordinates alliances with political support groups, draws on anthropologists, biologists, and others as advisors, brings legal suits, and negotiates with Hydro-Québec. A regional framework of supporting agencies includes the Income Security Board, Cree Hunters and Trappers Association, Cree Board of Health, Cree School Board, Cree Local Management Board, Cree National Youth Council, and James Bay Cree Native Women's Association. "Cree society has changed from being a fragmented society of seven distinct village bands with little unity – their integration at a regional level having been through non-Cree agencies and non-Cree officials – into a regional society where the villages ... have close ties with each other and administer their own affairs through a Cree governmental structure, staffed largely by Cree. They have turned seven [now nine] 'home villages' into one 'homeland'" (Salisbury 1986, 8). To be sure, the social form of many of these developments is decidedly Western and in that sense not an extension of traditional ways. But taken collectively, these changes in political organization may function to reinforce the social cohesiveness of Cree communities.

Thus far the analysis has focused on the broad spectrum of forces restructuring Cree life as a whole. What of the specific impact – the forces and events that bear upon particular individuals at particular places? It has already been mentioned that the entire community of Chisasibi was moved in the late 1970s from an estuary island at the mouth of the La Grande River to the mainland near by, while waterway diversions farther south left the village of Eastmain on a river that had lost ninety percent of its flow.

The La Grande project affected approximately half of the sixty traplines associated with the Chisasibi and Wemindji bands. If the Grande Baleine reservoir system is undertaken at a future date, about five percent of the total area of Whapmagoostui traplines would experience an impact (Hydro-Québec 1993). As well, twenty traplines in the Grande Baleine territory would be linked by new roads – an outcome the Cree have generally found advantageous (see Hydro-Québec 1993, 184, for the data). Three new airfields are also proposed for the Grande Baleine phase of development. One is a relocation of the existing Whapmagoostui-Kuujjuarapik airstrip as a noise-abatement measure. During the peak period of construction that airstrip would generate 25,000 flights per year, including such planes as the Boeing 737 (Hydro-Québec 1993).

Also significant among specific impacts is the loss of "special places," locations that carry particular social significance, ranging from archaeological burial sites to waterfalls that are aesthetically valued. In its summary environmental and social impact report on the Grande Baleine project, Hydro-Québec simply stated that "numerous" archaeological sites had been identified and that appropriate cartographic and photographic surveys, inventories, and digs would be conducted before construction was undertaken. Environmental and social impact studies, in general, have not found an adequate methodology for determining social worth. The language of Hydro-Québec's report is characteristic of a tendency to take inventory of specific impacts, to propose "mitigating measures" where possible (relocate the airport) without seeking a framework for aggregating these separate effects into a larger perspective. In fairness, such frameworks are still exploratory. We have reached the boundary line between what some call "objective" economic or scientific analysis and political-choice theories of valuation and decision making. (See, for example, Stevens's 1993 *The Economics of Collective Choice.*). For an interesting attempt to aggregate social worth over several categories of impact see the article "Valuation and Management of Wetland Ecosystems" in *Ecological Economics* (Costanza et al. 1989). Hydro-Québec acknowledges these methodological difficulties: "The financial evaluation of sociocultural impacts is even more difficult. Many believe economists do not have the right to put a price tag on goods or values for which it is difficult, if not impossible, to imagine a market. Such people may find it reprehensible that economists perform economic assessments of certain aspects of Native culture" (Hydro-Québec 1993, 175). One is left with simple declarations of overall impact. In its conclusion to the Grande Baleine study Hydro-Québec states:

The Grande Baleine complex will not cause an unacceptable impact: it will create a new natural environment that will prove to be productive.

The Grande Baleine complex will require the affected Native populations to adapt some of their wildlife harvesting activities and to change land use patterns in certain areas. For these populations, the complex will represent an additional cause of social change in the context of a significant and ongoing social transformation experienced over the last four decades. The complex will also provide Native peoples with an opportunity for economic development, enabling them to continue the transition to a wage economy while maintaining their traditional wildlife harvesting activities. (Ibid., 280).

To soften the long-term impact, Hydro-Québec advances a psychological perceptual argument, couched in cross-generational terms:

Of the Native peoples affected by the Grande-Baleine project, it is the current generation that will have to endure and adapt to the changes caused by the construction and eventual operation of the complex. Part of the current population has witnessed first-hand the major transformations that have taken place since the 1950s in Native communities. The prospect of further change, even though it was announced as far back as 1976, *may nevertheless push some to the limits of their ability to accept and adapt.* (Ibid., 278; emphasis added)

Placing native people (or humans in general) in an ecological context and then combining social impact analysis with environmental impact analysis has a certain holistic appeal. But the human dimensions of consciousness, agency, rights, perceptual relativism, and existential choice confound such efforts to find a seamlessness with Nature in a common biosphere. An alternative, albeit more circumscribed, perspective for considering the relationship of hydroelectric development to native peoples will be suggested below.

As part of the quest for a stronger methodological approach to environmental and social impact analysis, the argument has been advanced that higher recognition should be given to what is called "traditional ecological knowledge." Western scientific method is not the only route to understanding complex systems; indigenous peoples construct an analysis of their environment that is based on close observation, inferential statements of causality, and a particular nomenclature that should not be discounted out of hand. Moreover, "traditional ecological knowledge" may proceed from different foundational values concerning the existence, use, and meaning of the natural world. A workshop comprising twenty Australian and Canadian anthropologists and ecologists met in Canberra in 1988 to explore the significance of this issue and produced the volume *Traditional Ecological Knowledge: Wisdom for Sustainable Development* (Williams and Baines, eds. 1988). It argues, among other things, that traditional ecological knowledge is neither more nor less privileged than other scientific methods, that it is usually complementary to, rather than competitive with, western science, and that it is often a valuable source of knowledge that should be recognized and used. While the present study has benefited from numerous conversations with individual Cree leaders, hunters, teachers, and administrators, it falls far short of the systematic field research that would be needed to represent a full ecological understanding of the Cree.

Throughout the public airing of the James Bay controversy, the principal focus has been on dams and reservoirs as the instruments of disruption. Yet from a social impact perspective, the 1,500 kilometres of new roads traversing the region are the stronger agent of change. Historically, the geographic isolation of all but the most southern Cree villages sustained a social separation from French Quebec society that carried, if not the political meaning of autonomy, something of the reality of its practice (see chapter 6). To be sure, changes in schooling and health care, the world decline in fur prices, and the arrival of television had their separate impacts. But the coming of roads brought not only the subculture of motor vehicle ownership and travel but a new spatial order of interpenetration, exchange relationships, and social friction. The initial 620-kilometre road that paved the way for the La Grande system was accompanied by a construction boomtown, Radisson, not far from Chisasibi, serving over 20,000 construction workers. One study estimates that in 1990 over 25,000 tourists and sports hunters and fishermen travelled the La Grande road (Norbet 1991). A high proportion were local visitors from towns on the southern edge of the region. Of those classified as non-local, twenty percent "said they experienced mechanical breakdowns or emergency situations ... 50 percent [said] they knew little or nothing about the exclusive rights of James Bay Native people ... 70 percent were first-time visitors, and 80 percent said they would come back" (ibid.). Among all visitors, fifty-eight percent said they were entering the region to fish, and together they logged 53,000 fishing-days of effort (ibid.).

The roads have also brought increased hunting pressure. For some species, such as moose and bear, visitor competition with the Cree reduces the native hunt. Caribou hunting by non-natives, both by road and with fly-in outfitters, has produced its own story. Ever since 1984, when Hydro-Québec was accused of causing the drowning

death of ten thousand caribou as a consequence of opening a spill-
way on the Caniapiscau Reservoir (a charge they strongly dispute),
the circumstance of caribou in the region has received special attention (see (Hydro-Québec 1994, for example). Not well understood is
the fact that the present herd of about 800,000 (the largest in the
world) is in a long cycle of expansion from its 1951 level of 5000
members and is expected to crash. From this herd the Cree, Inuit,
and Naskapi take about 10,000 animals per year, while non-native
hunters take another 22,000 (interview, wildlife biologist Jean
Doucet, 6 September 1994). Whatever the cause of the 1984 drowning incident, the yearly kill of 32,000 caribou is the more salient
number – and the figure is growing – as access to the region
improves. But it too is small in the context of the natural population
collapse that is anticipated.

Once the trunkline roads had been built, the Cree themselves
actively sought linkages with other communities, including the coastal
villages of Eastmain and Wemindji and the inland community of
Nimaska. If the Grande Baleine reservoir is constructed, the road to
Radisson would be extended another 281 kilometres north, connecting Whapmagoostui and Kuujjuarapik. On the one hand the Cree
make significant use of the roads, shopping in commercial centres to
the south, travelling in and out of trapline areas more frequently and
easily, and visiting friends and relatives in other communities. But
social frictions have resulted as well. In the Cree magazine *The Nation,*
one article was titled "Sports Fishers Swarming up Route du Nord"
(Webb and Roslin 1994). Reports of native fishing nets cut by boat
motors or deliberately destroyed are joined by complaints of moose or
other game being shot in areas the Cree have traditionally regarded as
theirs, even though the 1975 agreement designates much of this territory as Category Three lands open to non-native use. In some cases
the friction arises between Cree and Cree, as casual "corridor" hunting
along roads by some encroaches on the traditional hunting grounds of
others. The James Bay Municipality, using funds from Hydro-Québec,
encourages tourist use of the region, constructing roadside rest areas,
boat ramps, service points, emergency sos telephones, and a tourist
information centre. At Radisson one finds garages, restaurants, and
accommodation, as well as guided tours of the second La Grande
power station.

In search of rapprochement with the Cree, Quebec, granted a $73
million contract for the new Nemaska-Chibougamau road to the
Cree Construction Company. The company was permitted to
conduct the environmental and social impact analysis, consulting
tallymen and trappers and securing state authorizations. (In one

instance, Cree trapper Freddy Jolly has argued that his trapline was significantly damaged by the road; seeking financial compensation, he has set up the "R-21 Trapline Defence Fund"; Nicholls 1994.) The Nimaska road, which serves a number of Hydro-Québec needs – paralleling powerlines, shortening the distance for large trucking contractors to Radisson, and well positioned for future development of the Nottoway-Rupert-Broadback river system – has nevertheless avoided all the controversy of the original La Grande road. Since the road was not proposed and constructed by Hydro-Québec but was sponsored by the Quebec Department of Energy and Resources and built by the Cree, the Nimaska route has been free of political turmoil.

The roads also lay down the infrastructure for other forms of economic development. Forestry operations are extensive in the southern portion of the region, enhanced by the La Grande and Nimaska roads. As one travels the southern third of the La Grande road, for example, great stretches of clearcut forest can be seen on either side. While there is great concern over rainforest cutting in equatorial nations, far less is said of clearcutting in boreal forests that border the northern taiga. It takes one hundred years for a black spruce to reach cutting size for a paper mill. While Quebec and other provinces establish mandates for replanting, continued "nutrient transfer" (or biomass removal) from thin soil on Canadian shield rock raises long-term questions of sustainability. In addition to encouraging forest cutting, roads are extending mining – historically a primary activity in the neighbouring towns of Val d'Or, Amos, and Chibougamau – into the James Bay region.

In all their ramifications roads accelerate the patterns of change that are likely to occur for "encapsulated" first nations, reducing the intergenerational time available to shape a new "transformational model," to use the language of the anthropologists. Such a model will find that cultural forms, social customs, and spiritual beliefs remain distinctly Cree, and that the Cree relationship to the land of the James Bay region continues to be strong. But the spring and fall goose hunts, expeditionary hunts for moose, bear, and caribou, or fishing for sturgeon will become supplementary activities, organized from a permanent residence in a village and woven into a pattern of livelihood derived from wage employment and transfer payments. The seventy-five percent who already conform to this pattern will become eighty, then ninety percent. For the fifty percent of Cree who are now under the age of twenty and the generation that follows them, the number of alternative opportunities for employment within the nine Cree communities may not keep pace with population growth. In that circumstance, educational and other paths may lead to or require out-migration to other parts of Canada.*

One of the recurrent criticisms of hydroelectric development in the James Bay region is that each phase of development is evaluated separately and independently, without a cumulative impact study. Concerns are voiced about changes in James Bay itself, in such areas as salinity, water circulation patterns, staging areas for migratory waterfowl, or esturine habitats for beluga whales. (See, for example, Hazell 1991, and Raphals 1992; for response, see Hydro-Québec 1993 pp. 139-147, 266-269) Yet the cumulative impact of increased human encroachment may be the greater concern, lead by the roads and compounding in its multiple, rippling effects. That, in turn, raises the fundamental question of who is gatekeeper-a question that forms the bridge to the next and final step in this analysis.

Ecological Injury or Constitutional Standing?

The opposition of Cree leadership to hydroelectric development in this large northern territory has been formulated and voiced primarily in terms of environmental and social impacts. An alternative approach to understanding the relationship of indigenous societies to the economic development agendas of encapsulating nation states is to frame the analysis within concepts of autonomy, self-determination, and constitutional standing. In what measure is it possible to construct a new architecture of self-determination for these societies that is distinct from the model of the sovereign nation state but preserves a number of key linkages to the societies that surround them?

The paramount focus on environmental and social impact that has shaped the debate thus far has been encouraged by a number of organizations (largely in the us) that have made common cause with the Cree and the James Bay controversy. The Audubon Society, Greenpeace, and Friends of the Earth have mounted major campaigns. The Natural Resources Defense Council, in a public mailing urging political action and soliciting donations, stated in boldface that the James Bay project represented: "construction so monumental it will cause

---

*As I travelled up the La Grande road with biologist Raymond Coppinger, we pulled over to see if two people standing by a pickup truck needed assistance. A young Cree in his twenties thanked us and explained they were on their way to try to shoot a moose, using the all-terrain vehicle that was sitting in the back of the truck. He was in his third year of college, majoring in accounting. They had stopped so that his friend could remove the quills from a roadkill porcupine, to take them, he said, to an old woman in Waskaganish for use as clothing decoration. That vignette in many ways captures the cusp between old and new.

wildlife and wilderness destruction on a scale never before witnessed in North America." (For an examination of environmental impact that provides an analytical framework and reaches a rather different conclusion, see chapter 3. For an analysis of the strategic choices of the Cree, as understood in a historical and cultural context, see chapter 6.)

The issue, of course, is not an oppositional choice between environmental and social impact analysis on the one hand and a constitutional-standing argument on the other. But the two approaches entail quite different first premises, burdens of proof, and patterns of supporting evidence. The analysis in the sections above has sought to place the impact of hydroelectric development within a larger understanding of the on-going forces – demographic, social, and economic – that are reshaping Cree life.

What, then, does it mean to construct a position around the concepts of autonomy, self-determination, and constitutional standing? Among the points of departure are such questions as whose lands are these? Who shall be gatekeeper? And on what terms shall economic development take place if it proceeds at all? For some, these questions seek to reopen a closed book. In signing the James Bay agreement of 1975 the Cree explicitly relinquished all historical aboriginal claims and agreed to a three-tiered classification of land that permits hydroelectric development and regards the largest part of the region to be public territory. The Canadian Federal Court of Appeal ruled in 1992 that the Cree understood what they were signing, and that "the principle that ambiguities must be construed in favour of the Aboriginals rests, in the case of historic treaties, on the unique vulnerability of the aboriginal parties, who were not educated and were compelled to negotiate with parties who had a superior bargaining position ... In this case there was simply no such vulnerability" (Federal Court of Appeal, Nov. 1992). But the international standing of indigenous, or "first," peoples is anything but a closed book. The issue has always been a political one, revisited in successive generations within a set of political and power relations specific to each.

While the path is strewn with broken treaties and redefined territories, the present reconsideration has reversed the grain, raising fundamental questions about the basic rights and claims of indigenous peoples. Interesting lines of thought are probing the dimensions of nationhood and conceivable configurations of "nations within nations." How can autonomy and self-determination be advanced while other elements held in common with the encapsulating society – currency, security, social transfer payments, infrastructural support – remain in place?

The movement towards First Nations within nations has been slow and exploratory. In 1957 the International Labor Organization issued Convention 107 on "Protection and Integration of Indigenous and other Tribal and Semi-Tribal Populations in Independent Countries," a convention that various countries have since adopted or endorsed. It affirmed a basic right of indigenous ownership, collective or individual, over traditionally occupied lands. But it reserved to the surrounding nation-state the authority to take these lands "for reasons relating to national security, or in the interest of national economic development or of the health of the said populations" (Morse and Berger 1992, 18). In the case of such a taking the convention called for compensation and the provision of additional lands of at least equal quality-a framework that evades matters of cultural identity and autonomy.

Also instructive has been the World Bank's response to criticisms that its support of large-scale development projects has often been at the expense of indigenous peoples. In 1980 the World Bank announced a new policy on "Social Issues Associated with Involuntary Resettlement in Bank-Financed Projects" (World Bank 1980, n.p.). While the new policy accepts the notion of involuntary resettlement, the Bank is conflicted in its recognition of "the human suffering and hardship caused by involuntary resettlement," the difficulty of valuing "non-economic assets," and the complexity of trying to preserve "cultural integrity." Without yielding on the idea of properly handled involuntary resettlement, the World Bank's policy statement nevertheless declares:

Experience has shown that, unless special measures are adopted, tribal people are more likely to be harmed than helped by development projects that are intended for beneficiaries other than themselves ... [H]ow can the government harmonize its interest in the development of a rich ore body or a major hydro potential with the need to safeguard the rights of tribal people in the project area? These are matters for judgements guided by the principle that Bank assistance should help prevent or mitigate harm, and provide adequate time and conditions for acculturation. (Ibid., n.p.)

The World Bank's own words came back to challenge it in 1992. With controversy growing over its support for the large-scale Sardar Sarovar Project to dam India's Narmada River, the Bank appointed an independent review panel to evaluate the environmental and human impacts. The panel concluded that "engineering and economic imperatives have driven the Projects to the exclusion of human and environmental concerns" and that "involuntary resettlement from the

Sardar Sarovar Projects offends recognized norms of human rights – human rights that India and the Bank have been in the forefront to secure" (Morse and Berger 1992, xxiv, xx). The World Bank withdrew its support for the Narmada project pending reassessment and changes; the Indian government decided to continue construction without the bank's funds.

Within Canada the political autonomy and constitutional standing of native peoples is at one of the most fluid points of reconsideration to present itself in the past century. In the ongoing constitutional crisis between French- and English-speaking Canada, aboriginal people have emerged as a major third force. (For a historical expansion on this struggle, see Pawson and Cant 1992; Usher et al. 1992; Waldram 1987.) The efforts to accommodate francophone Canada within a new federalism that recognizes Quebec as a "distinct society" has brought a parallel demand from native societies for a comparable reassessment. The constitutional compromise proposed in the Meech Lake Accord unravelled in no small part because of the protests of native peoples, particularly in Manitoba.

In the meantime, province by province, new agreements with indigenous societies are being crafted. In 1991 the Alberta government adopted measures to guarantee native peoples both a land base and greater local control. But Rosemarie Kuptana, president of the Inuit Tapirisat, of Canada questioned the adequacy of this step: "Entrenching only a right to delegated municipal-like powers would mean recognizing a right flowing not from the humanity of indigenous peoples, but from the discretionary goodwill of governments" (in Cernetig et al. 1991, n.p.).

In Ontario a more fundamental agreement was reached in 1991. Titled "A Statement of Political Relationship," it recognized a pre-existing right of self-government that derives from historical roots – a right that "flows from the Creator and from the First Nations' original occupation of the land" (Mackie 1991, n.p.). The Ontario agreement was explicit in recognizing an "inherent right" of superior standing to the delegated rights of municipal governments. At the same time the agreement reaffirmed the obligation of the federal government to provide services to native communities, while also noting, in the words of Ontario's minister for Native Affairs, that "we are dealing with a self-governing community that has a right to decide how and when and why [government] services will be delivered" (ibid., n.p.). But Ontario's regional representative to the Assembly of First Nations (AFN), Chief Gordon Peters, pressed the point that past practices of financial compensation for natural resource use have not been sufficient, and that conflicts over land use in mining, logging, fishing, and hunting will have to be more directly addressed, if the agreement is to have any force (ibid., n.p.).

As the constitutional debate intensified and the possibility of a divided Canada impelled the search for a new constitutional foundation, the federal government appointed a Royal Commission on Aboriginal Peoples in April 1992. One of the most expensive such commissions in Canada's history (and marred by internal dissent), it held hearings in over one hundred native communities. In one of its reports, *Focusing the Dialogue* (n.d.), the commission set forth in broad terms the twin goals of territorial control and expanded self-government. In 1996 the Canadian government formally apologized for the injustices and racism experienced by native peoples and established a multimillion-dollar fund for the payment of compensation.

Parallel to these efforts, Canada's Assembly of First Nations has established its own constitutional commission. Ovide Mercredi, former national chief of the AFN, advanced four major goals for constitutional reform: "constitutional recognition of natives as distinct, the right to native self-government, full and equal participation in reforming the Constitution, and recognition the federal and provincial governments have limited jurisdiction over native affairs" (Cernetig et al. 1991, n.p.). Viola Robinson, speaking as president of the Native Council of Canada, underscored these objectives. In contrast to the belated recognition of indigenous peoples' concerns in the Meech Lake negotiations, "we will be full and central participants in this round" of constitutional talks (ibid., n.p.).

Within Quebec, the call for territorial self-determination for indigenous people confronts directly the central role of hydroelectric development in Quebec's long-range economic agenda. In 1991 the then premier, Robert Bourassa, expressed his displeasure: "The natives blocked the ratification of the Meech Lake Accord. Now should they block Great Whale? There are seven million Quebeckers on this land, in Canada. They are an important group" (Seguin 1991, n.p.).

In 1994 the return to power, with a slim majority, of the Parti Québécois brought with it the promise of a referendum on Quebec separation and independence – a referendum that the separatists subsequently lost by a very narrow margin. It also brought continuing uncertainty about the constitutional status of Quebec's indigenous peoples and provided an intriguing political opening for the Cree. As noted, Canada's Federal Court of Appeal ruled in 1992 that when the Cree signed the 1975 James Bay agreement, they knowingly relinquished all prior aboriginal claims in exchange for this modern settlement. Nevertheless, the unprecedented possibility of the separation of Quebec has created a context in which the 1992 ruling might well be held moot on the grounds that it is superseded by the issue of the constitutional standing of Quebec's indigenous people *vis-à-vis* the rest of Canada. Coterminous with the resolution of the debate on the constitutional status of the Cree, Inuit, and others is the issue of the constitutional status of the lands of the vast northern territory. Any effort to retain this territory as federal land on the grounds that federal jurisdiction supersedes provincial jurisdiction in Indian affairs would present a serious challenge to the economic integrity of a newly formed Quebec nation. The issue could easily become a sticking point that would work its way through international forums and tribunals.

No surprise, then, that the newly elected Quebec premier, Jacques Parizeau, made overtures towards negotiating a new understanding of native self-government and held out the possibility of royalty payments for natural resources extracted from native lands (Thompson 1994). Cree Grand Chief Matthew Coon-Come indicated a willingness to discuss such matters but cautioned that parts of existing agreements, in such areas as education and policing, had not yet been fully met. Speaking for the Assembly of First Nations, Ovide Mercredi made clear the challenge that would be presented in the event of Quebec's separation: "The First Nations in Quebec have the same rights to self-determination that are being asserted by the sovereignists in Quebec. If Canada is divisible, so is Quebec" (Thompson 1994, n.p.). Two weeks later Matthew Coon-Come withdrew from a public ceremony to sign a new policing agreement on the grounds that Premier Parizeau and his government had not initiated formal relations with the Cree nation from a cabinet-level ministry. The Cree position was more forcefully delineated in a "Resolution on Cree Status and Rights" introduced at the 1994 Annual General Assembly. It states, in part:

Whereas we, the Cree People (Eenouch), have governed ourselves, and owned, occupied, and been sustained by our traditional lands and waters, including offshore, for thousands of years; ...

Whereas it is the stated core policy of the Quebec and federal separatist parties that the Quebec people have the right to self-determination, and that pursuant to this right a separatist government will immediately commence steps to achieve the separation of Quebec from Canada; ...

Whereas any secession of Quebec from Canada that purported to unilaterally remove the Crees and their traditional territory from the Canadian federation would constitute a fundamental alteration to and repudiation of the James Bay and Northern Quebec Agreement, 1975;

Whereas rights accruing to the Crees under the terms of the James Bay and Northern Quebec Agreement are constitutionally guaranteed and affirmed as treaty rights, pursuant to section 35 (1) of the Constitution Act, 1982;

Be it resolved:

That the Cree Nation reaffirms its fundamental status and rights in the context of the possible secession of Quebec from Canada, including our right to self-determination and to freely determine our political status and that of our traditional territory. (*The Nation* (Cree), 22 September 1994)

Such are the multiple grounds on which the Cree would, paradoxically, use their federal and constitutional status as a protected minority to advance the case for a distinct, separate, self-governing society.

Conclusion

Contrary to the portrait that has emerged from political action groups and parts of the public press, hydroelectric development in the James Bay region is not the primary transformational force reshaping the lives of the Cree people. It is true that very specific impacts have displaced particular individuals and altered family patterns of land use. Yet these changes do not represent the defining or essential forces at work. Indeed, if any single part of the hydroelectric development story were singled out for its contribution to social change, it would be the role of the income security program in sustaining a key component of traditional Cree society. During the more than two decades since development began, changes in culture, livelihood, community infrastructure, and political organization have proceeded in parallel fashion in Cree communities within the path and without the path of hydroelectric development.

To the extent that hydroelectric projects have contributed to the totality of change experienced by the Cree people, it is the network of roads – more than the dams and reservoirs – that encroaches upon the social distance and patterns of interchange between Cree and non-Cree.

Both internationally and within Canada the great historical question of the standing of first nations within encapsulating nations repeats itself with new ferment, seeking within its language a stronger basis for advancing the principles of autonomy and territorial control. At this particular historical moment, the self-governance movement gains leverage both from a greater measure of pan-tribal political organization and from a broader base of empathy and political support within the larger metropolitan society. To move from abstract generalities into a specific agenda for change will require negotiations at a sharply defined level that includes issues of employment, schooling, policing, resource management, judicial authority, federal transfers, and infrastructural design, among others.

At the heart of Cree social identity is the historic connection to the land, the collective experience of the hunt, the seasonal rhythms of dispersion and recombination, and the transfer from elder to younger generations of the spiritual and social legacy of the land. Yet as year-round residence within the nine Cree communities expands to three-fourths of the population, the next generation, in particular, faces unresolved questions about new forms of livelihood and meaningful work. Within the political framework of self-determination, the issue of seeking to block hydroelectric power or other development proposals by outside interests provides the particular arena or context for pursuing an expansion of autonomy and indigenous control. But the more difficult task remains – the task of speaking to the development needs of the present generation of Cree people without compromising the continuity of collective self so strongly advanced as the foundation of native identity.

REFERENCES

Abu-Lughod, L. 1986. *Veiled Sentiments: Honour and Poetry in a Bedouin Society.* Berkeley, CA: University of California Press.

Beauchemin, G. 1992. "The Unknown James Bay and Northern Quebec Agreement." *Forces*, no. 97 (Spring):29–35.

Beaulieu, D. 1984. *The Crees and Naskapis of Quebec.* Quebec City, PQ: Gouvernment du Quebec, Ministère de l'Industrie et du Commerce.

Bender, B., and B. Morris. 1991. "Twenty Years of History, Evolution and Social Change in Gatherer-Hunter Studies." In *Hunters and Gatherers: Property, Power and Ideology*, edited by T. Ingold, D. Riches, and J. Woodburn. New York: Berg Publishers.

Cernetig, M., S. Delacourt, and D. Wilson. 1991. "Native Self-Government Accepted by Premiers." *Globe and Mail*, Toronto, 27 August 1991.

Coon-Come, M. 1991. "Where Can You Buy a River?" *Northeast Indian Quarterly* 8, no. 4:6–11.

Cornell, S. 1988. *The Return of the Native: American Indian Political Resurgence.* New York: Oxford University Press.

Costanza, R., S. Farber, and J. Maxwell. 1989. "Valuation and Management of Wetland Ecosystems." *Ecological Economics* 1:335–61.

Cree Hunters and Trappers Income Security Board. 1993. "Annual Report." Sainte-Foy, Quebec.

Deloria Jr., V., and C.M. Lytle. 1984. *The Nations Within: The Past and Future of American Indian Sovereignty.* New York: Pantheon Books.

Feit, H. 1979. "Political Articulations of Hunters to the State: Means of Resisting Threats to Subsistence Production in the James Bay and Northern Quebec Agreement." *Études/Inuit/Studies* 3, no. 2:37–52.

– 1980. "Negotiating Recognition of Aboriginal Rights: History, Strategies and Reactions to the James Bay and Northern Quebec Agreement." *Canadian Journal of Anthropology* 1, no. 2:159–72.

– 1982. "The Future of Hunters Within Nation-States: Anthropology and the James Bay Cree." In *Politics and History in Band Societies*, edited by B. Leacock and B. Lee. Cambridge: Cambridge University Press.

– 1989. "James Bay Cree Self-Governance and Land Management." In *We Are Here: Politics of Aboriginal Land Tenure*, edited by E. Wilmsen. Berkeley, CA: University of California Press.

Grodinsky, W.S. 1991. "The James Bay and Northern Quebec Agreement." *Northeast Indian Quarterly* 8, no. 4:48–51.

Hazell, S. 1991. "Environmental Impacts of Hydro-Development in the James Bay Region." *Northeast Indian Quarterly* 8, no. 4:20–2.

Hydro-Québec. 1993. "Grande-Baleine Complex Feasibility Study – Summary." Montreal: Hydro-Québec.

– 1994. "The Drowning of 9,604 Caribou." Discussion Paper No. 5. Montreal: Hydro-Québec.

Ingold, T. 1991. "Notes on the Foraging Mode of Production." In *Hunters and Gatherers: Property, Power and Ideology*, edited by T. Ingold, D. Riches, and J. Woodburn. New York: Berg Publishers.

LaDuke, W. 1991. "Environmental Work: An Indigenous Perspective." *Northeast Indian Quarterly* 8, no. 4:16–19.

Leacock, E. 1954. "The Montagnais Hunting Territory and the Fur Trade: American Anthropological Association." *Memoir* 78, no. 56:Pt 2.

Mackie, R. 1991. "Ontario to Recognize Native Governments." *Globe and Mail*, Toronto, 6 August 1991.

Masty, Sr., D. 1991. "Traditional Use of Fish and Other Resources of the Great Whale River Region." *Northeast Indian Quarterly* 8, no. 4:12–15.

Morse, B., and T. Berger. 1992. *Sardar Sarovar: Report of the Independent Review.* Ottawa: Resources Futures International.

Murphy, R., and J. Steward. 1956. "Tappers and Trappers: Parallel Process in Acculturation." *Economic Development and Culture Change* 4:335–55.

Nicholls, W. 1994. "It's Time to Listen to the Trappers." *The Nation* (Cree), 17 June 1994.

Norbet, M. 1991. "Recreational and Tourist Use of Roads in the James Bay Region." Paper presented at the ACFAS Conference, Sherbrooke, Quebec, May.

Pawson, E., and G. Cant. 1992. "Land rights in historical and contemporary context." *Applied Geography* 12:95–108.

Peters, E. 1992. "Protecting the Land under Modern Land Claims Agreements: The Effectiveness of the Environmental Regime Negotiated by the James Bay Cree in the James Bay and Northern Quebec Agreement." *Applied Geography* 12:133–45.

Pothier, R. 1967. "Relations inter-ethniques et acculturation à Mistassini." *Travaux divers*, no. 9. Quebec: Centre d'Études Nordiques, Université Laval.

Raphals, P. 1992. "The Hidden Cost of Canada's Cheap Power." *New Scientist* (February 15): n.p.

Richardson, B. 1991. *Strangers Devour the Land*. Post Mills, VT: Chelsea Green Publishing.

Roy, L., S. Beaudet, and D. Vandal. 1991. "Harvesting of Wildlife Resources in the James Bay Region." Paper presented at the ACFAS Conference, Sherbrooke, Quebec, May. Quebec: Hydro-Québec/Ministère du Loisir de la Chasse et de la Pêche.

Salisbury, R. 1986. *A Homeland for the Cree: Regional Development in James Bay 1971–1981*. Montreal and Kingston: McGill-Queen's University Press.

Scott, C. 1983. "The Semiotics of Material Life Among Wemindji Cree Hunters." Ph.D. dissertation. Montreal, McGill University.

– 1984. "Between 'Original Affluence' and Consumer Affluence: Domestic Production and Guaranteed Income for James Bay Cree Hunters." In *Affluence and Cultural Survival (Proceedings of the American Ethnological Society, 1981)*, edited by R. Salisbury and E. Tooker. Washington, DC: American Ethnological Society.

– 1988. "Property, Practice and Aboriginal Rights among Quebec Cree Hunters." In *Hunters and Gatherers: Property, Power and Ideology*, edited by T. Ingold, D. Riches, and J. Woodburn. New York: Berg Publishers.

– 1989. "Knowledge Construction Among Cree Hunters: Metaphors and Literal Understanding." *Journal de la Société des Américanistes* 75:193–208.

Seguin, R. 1991. "Hydro Delay Angers Bourassa." *Globe and Mail*, Toronto, 5 August 1991.

Senecal, P., and D. Egre. N.d. "Impacts of James Bay Projects on Cree Communities." Unpublished paper. Hydro-Quebec.

Stevens, J. 1993. *The Economics of Collective Choice*. Boulder, CO: Westwiew Press.

Tanner, A. 1979. *Bringing Home Animals: Religious Ideology and Mode of Production of the Mistassini Cree Hunters*. St John's: Memorial University of Newfoundland, Institute of Social and Economic Research.

– 1983. "Introduction: Canadian Indians and the Politics of Dependency." In *The Politics of Indianness: Case Studies in Native Ethnopolitics in Canada*, edited by A. Tanner. St John's: Institute of Social and Economic Research.

Thompson, E. 1994. "Native Leader Skeptical about PQ Granting Rights." *The Gazette*, Montreal, 13 October 1994.

Turgeon, P. 1992. *Radissonia: The James Bay Adventure*. Translated by Dominique Clift. N.p.: Libre Expression.

Usher, P., F. Tough, and R. Galois. 1992. "Reclaiming the Land: Aboriginal Title, Treaty Rights and Land Claims in Canada." *Applied Geography* 12:109–132.

Waldman, C. 1984. *Atlas of the North American Indian*. New York: Facts On File.

Waldram, J. 1987. "Relocation, Consolidation, and Settlement Pattern in the Canadian Subarctic." *Human Ecology* 15, no. 2:117–31.

Waldram, J.B. 1988. *As Long as the Rivers Shall Run; Hydroelectric Development and Native Communities in Western Canada.* Winnipeg: University of Manitoba Press.

Webb, E., and A. Roslin. 1994. "Sports Fishers Swarming up Route de Nord." *The Nation* (Cree), 1 July 1994.

Williams, N., and G. Baines, eds. 1988. *Traditional Ecological Knowledge: Wisdom for Sustainable Development.* Canberra, Australia: Centre for Resource and Environmental Studies, Australian National University.

World Bank. 1980. "Operation Manual Statement No. 2.33." In B. Morse and T. Berger, *Sardar Sarovar: Report of the Independent Review.* Ottawa: Resources Futures International. (1992).

6 Culture, Social Change, and Cree Opposition to the James Bay Hydroelectric Development

ADRIAN TANNER

Introduction

This chapter examines Cree opposition to the proposed Great Whale hydro project. Obviously, the project represented different things to different people. For many Québéquois, it was both a major potential source of employment and an important contribution to Quebec's future nationalist (but not necessarily separatist) economic development strategy. For environmentalists it represented a major alteration to a complex and delicate subarctic ecosystem, with several potentially serious impacts. But for the aboriginal peoples of the area, the Cree and the Inuit, the project did not just involve the alteration of the area's physical environment: it would introduce major social changes to the region over which they would have little or no control. These social changes could well have severe negative consequences, including, it was feared, social breakdown in their communities and the consequent loss of control over, and deterioration of, their way of life.

To arrive at a reasonable judgment about the validity of these kinds of concerns is not a trivial exercise. Experts and stakeholders disagree about the social impacts that follow on the heels of such large projects, particularly one that would open up the first road connection to the area and bring in thousands of workers and other outsiders. In the case of the Great Whale River proposal, the Cree community of Whapmagoostui saw themselves as having avoided the negative social changes that had disrupted some of the Cree communities to the south, changes that, in their perception, flowed from the La Grande project, the first phase of the James Bay development, begun in 1971 (Niezen 1998). Unfortunately, Hydro-Québec, the provincial government utility most directly involved in the project, took the position that concern over social impact was beyond the utility's area of responsibility, and that no social issues associated with the project would be allowed to influence the decision whether or not to go ahead with the development. Nor did any other public agency, beyond those of the Cree and Inuit, step in to address the issue. There was little public discussion about the project's potential social impacts, outside the Cree and Inuit communities themselves. Instead, both the Cree and Hydro-Québec launched intensive lobbying and publicity campaigns as they battled over the fate of the project. The Inuit were internally divided on the issue, some joining with the Cree opposition, others negotiating for financial benefits from the developer.

Before getting down to the substance of the chapter, two limitations should be noted. First, the information presented here is based on public positions taken by Cree leaders, and on my general knowledge of the Cree of Quebec (I have been involved in research among the Cree for many years, including several weeks of fieldwork in 1977 and 1990 in the Cree community of Whapmagoostui at Great Whale River). This chapter does not examine the predictions of the Cree and others about the impact of the Great Whale project. It is more an outcome of my own long-term general interest in the Cree of Quebec. I also mention, in less detail, the position of the Inuit, in the belief that a general comparison between the Cree and Inuit cultures and the differing ways each has reacted to externally induced change can provide us with important insights, both into their common concerns as aboriginal groups undergoing change and their own distinctive perspectives on the development project. My comments on the Inuit are based on general knowledge and the literature, rather than on specific field study.

Second, although I focus on concerns about the social, as opposed to the physical, impacts of the project, it is not always easy to distinguish between the two. Whether a physical effect is considered to be positive, negative or neutral depends in part on the value attached by the members of a particular social group to the physical environment that is likely to be affected. Thus, a physical effect acquires a social dimension in the way it is understood and evaluated by a group. The Cree's view of the likely physical effects derive, in part, from their own culturally-framed knowledge of the physical environment, knowledge gained though long-term empirical observation. This knowledge is to be seen as parallel to, rather than identical with, Western scientific knowledge. Because it is based on intimate and highly practical experience of the environment on which the survival of the Cree depends, it is not a body of knowledge that can be dismissed as uninformed.

Because perceptions about the physical environment have this social aspect, the Cree's concerns about the social impacts of development must be seen in the context of their understanding of what the potential physical impacts might be. This understanding was based both on knowledge of the regions that would be involved and on knowledge, and in some cases direct experience, of the environmental effects of the now complete La Grande project. The Cree were concerned, for example, that flooding by the Great Whale project would affect their most productive harvesting lands, that is, the lowlands along the main waterways, as was the case in the equivalent areas flooded by the La Grande project. It is in these lowlands that most of their campsites, portages, buried ancestors – in short their memories of the land – happen to be located. There were also concerns about the proposed diversion of the lower portion of the Great Whale river, with a consequent reduction in downstream flow and the conversion of the old river mouth at the site of the Cree community of Whapmagoostui into a saltwater inlet. The Cree experience of the negative impact of the diversion of the lower portion of the Eastmain River, undertaken as part of the La Grande project, was significant in this regard.

There were other concerns about the impact of the proposed road construction over large parts of the territory, and particularly about the effect on game populations of increased public access to the region – concerns that elsewhere had come in the wake of the La Grande project. In particular, it was feared that, with the potential influx of sports hunters, certain game species, even those now abundant like caribou, might become scarce. Finally, there was concern over the potential impact of the new outlet to the Great Whale River. While the present river mouth discharges into the open waters of Hudson Bay, the new outlet, which would have an augmented flow, would discharge into Manitunik Sound, possibly changing the salinity of the water in the sound, or even the seasonal occurrence of freeze-up and break-up. The effect of such changes on this critical hunting area was a matter of speculation, but the concerns were serious. This was perhaps the single greatest point of trepidation for both the Cree and Inuit in terms of physical impacts, particularly since the La Grande project offered no comparable situation.

While these concerns were perfectly reasonable from the Cree point of view, Hydro-Québec and other government agencies seemed to think that to address these issues seriously in the Cree's own terms would have been an admission of defeat. Instead they chose to hide behind the Western bureaucratic process and scientific jargon of the environmental impact statement, which few if any members of the public (Cree or otherwise) could be expected to follow. I will not be dealing with the specifics of Cree environmental knowledge and concern, but it is important to acknowledge that their concerns were real; that they were held generally among the experienced hunters within the community, not just the political leaders; and that the issues were not manufactured for rhetorical purposes.

Social Impact Assessment

Most aboriginal communities throughout northern Canada have what may be referred to as the classic symptoms of a specific form of social breakdown, associated with the rapid disintegration of the traditional way of life, lack of satisfactory alternatives, and unsatisfactory adaptation to what few alternatives there are. Although alcohol, substance abuse, family violence, suicide, and other social pathologies are widespread in these communities, they are by no means found to the same extent in each. Some communities have been less affected, while others, after a difficult period, have succeeded in taking matters in hand. The worst cases often result from situations of especially rapid, unplanned, and externally induced social change, often involving urbanization, community relocation, and a major influx of newcomers. Such changes are often imposed without sensitivity to the local way of life and without the agreement or participation of the community, thus inducing a sense of powerlessness (for various cases that conform to this general pattern, see Goddard 1991; Shkilnyk 1985; Waldram 1988). Napoleon (1992) coined a term for the phenomenon – "Posttraumatic stress syndrome." As with many phenomena at the boundaries of scientific knowledge, specific conditions that would enable us to predict the occurrence or intensity of this form of social breakdown have yet to be pinpointed. Nor is current social scientific knowledge about development-related social breakdown well understood by the public at large.

Canadian environmental legislation is intended to ensure that objective scientific knowledge, including knowledge of anticipated negative social impacts, is used in development planning in a manner that is beneficial to the public interest. In practice, however, the contribution by the social sciences to the assessment process often tends to be discounted and marginalized. As a result, given that in cases involving social development, social science issues may also be highly political, decisions concerning these issues may be influenced by powerful political interests. Although social breakdown in Canadian aboriginal communities is widespread, it is an area in which agencies like Hydro-Québec have little expertise (but see Senécal and Égré 1998), and regarding which they are, quite naturally, reluctant to be held responsible. In the dispute over the proposed Great Whale hydroelectric project several very powerful political interests became involved, not only Hydro-Québec but national, provincial, and state governments, as well as the Grand Council of the Cree, and some Inuit organizations, and aboriginal rights and public interest environmentalist groups in Canada, the US, and Europe. As the public debate surrounding the environmental impact assessment became increasingly politicized and deintellectualized, the examination of arguments about social breakdown and other potential social impacts became a largely meaningless exercise. The more highly publicized debates that did take place involved claims and counterclaims about the underlying motives of the various interest groups in taking the positions they did over Great Whale.

Admittedly, non-Cree who try to understand and evaluate the validity of the Cree perspective face certain difficulties. In one particular "Catch 22," it emerged that much of the social impact research undertaken for the environmental impact assessment (Scott and Ettenger 1994) was never considered by the proponent, Hydro-Québec, partly because, they say, it was not received in time but also because it was largely based on Cree testimony. To the proponents it seemed to be biased, especially since the subjects of the research belonged to an organization that was on record as being politically opposed to the project. As a result the local aboriginal people, who would bear the brunt of any impact from the project, were once more effectively marginalized in the assessment process. Yet the testimonies yielded much factual data, some based on actual experience of the La Grande project, that could have been independently verified.

Still more problematic was the question how the assessment process, which was tailored to non-Cree modes of thinking, could comprehend and incorporate the framework of Cree knowledge – of their own territory, their own society, and more. How could the assessment panel take account of the significance and value to the Cree of aspects of their environment and social life that were likely to be affected by the project? The Cree's motives and strategies in opposing the project were difficult for the outsider to grasp, hence easier to dismiss out of hand.

As already noted, public policy decisions about projects like the proposed Great Whale development are today expected to pass an environmental and social impact assessment. This process uses the Western scientific knowledge framework in an attempt to offer a rational and objective aid to public decision making. However, experts seldom agree on what the outcomes are likely to be, and science can seldom tell us definitively which of two outcomes will be better for society. In other words, any claim that the scientific method can produce an authoritative, value-free, and culture-free understanding of the world is deceptive. Ambiguities inherent in the assessment process are multiplied when the major stakeholders involved in a proposed development from quite different cultural groups, each with its own conventions and cultural ideals. It is therefore becoming recognized that an indigenous knowledge perspective should be included in environmental impact assessments involving aboriginal groups (for some examples, see Nakashima 1990; Usher 1987; Waldram 1986; Werner 1980).

The Politics of Knowledge

Both Cree and non-Cree have distinctive understandings of the environment, animals and plants, human society, and potential strategies by which humans can deal with the impact of environmental change on human society. The Cree have well-developed practical ideas about such matters as the ecology of their region, resource fluctuations, the relationship of such changes to Cree spiritual entities (i.e., beings associated by the Cree with animals and the forces of nature), social development objectives, and human values. Without going into detail, Cree knowledge does not always accord with the Western approach. The general area of "wildlife management" is a case in point. According to many Cree hunters, the relationship between human beings and animals cannot be understood solely in terms of an ecosystem model that only looks at materialistic interchanges among life forms; for them, there are moral and spiritual aspects to be considered as well. Many Cree believe that game species are sensitive to any lack of "respect" shown by humans (Tanner 1979). Consequently, they are unwilling to accept that game populations can be "managed" merely for the benefit of humans without, at the same time, imposing certain moral constraints on the managers.

The public debate over Great Whale that took place in Canada and the US paid little attention to Cree knowledge. Cree spokespeople were heard from during the debate, but it was difficult, particularly in the context of an "event" largely staged for the media, for the distinctive Cree perspective to make itself heard. As a result the main opposition to the project has been couched in Western "environmentalist" terms without benefit of Cree concepts and perspectives.

As the parties to the debate failed to accommodate the diversity of perspectives on Great Whale, especially those arising between the aboriginal people and Hydro-Québec, the issue was reduced to one of political power, that is, whether and in what form the Cree and the Inuit were to exercise their interest in the development of their territory, despite other interests. Were they to have any significant influence over public policy making such that they could, to a considerable extent, determine the course of development, and in a form compatible with their own cultural understandings?

The suggestion that the Cree and Inuit world views should be taken into account in public policy decisions is itself an example of the much larger issue of self-determination now under debate in Canada. Can distinct cultural groups, including aboriginal peoples and, significantly, Québécois, be given the power to direct their own futures within the Canadian state rather than be directed by people of other cultural backgrounds and understandings? The recent report of the Royal Commission on Aboriginal Peoples (1966) spoke of their "inherent right to self-government." This has particular resonance in the Canadian North, where aboriginal people form a major part of the population, and where they effectively possess territory in the sense that most land, while claimed by the Crown, remains accessible to local aboriginal people.

It may be for largely pragmatic reasons that aboriginal self-government is receiving serious consideration now, beyond any legal arguments in its support, or any new public respect for alternate aboriginal forms of knowledge. Past attempts to integrate aboriginal people into mainstream Canadian society or to impose on them external Euro-Canadian forms of governance have not only failed but have also contributed to the creation of the intractable social pathologies that are found within aboriginal communities. As a way of dealing with these social problems, self-government is becoming recognized as potentially beneficial to Canadian society and aboriginal groups alike, particularly in the North.

The idea that Cree views about the development of their homelands ought to be taken into account is based on two principles. First, as the people who will be most directly affected by a development project, local Cree and Inuit concerns require attention. This "adjacency priority" principle is incorporated into current Canadian Environmental Impact Assessment procedures. Secondly, as indigenous people, they have a special interest in any proposed development of the aboriginal homeland that they occupy and, to varying degrees, live off. While the state also has an interest in the use of the region, the Cree would probably assert development should be undertaken only if it satisfies their interests as well as the interests of the state. That this principle may not have been followed in the past does not mean it is not a sound principle of public policy.

The Apparent Acceptance of Hydroelectric Development by the Cree

On the surface, it might seem as if there has been a significant turnaround in Cree thinking about hydroelectric development since 1974, when they undertook to negotiate with Quebec and Canada an agreement to allow the building of the La Grande project under specific conditions. The Cree took a lead role in this negotiation process, working actively and optimistically towards the set of objectives that had been set out at the start of the talks in an agreement-in-principle. The Cree have, moreover, made several subsequent agreements on hydroelectricity in their region, effectively accepting modifications to the original La Grande project. All this might suggest that the Cree were enthusiastic about the James Bay agreement and had reached an irrevocable accommodation with the hydroelectric development of their region. As we shall see, the suggestion is misleading.

The James Bay agreement represented, for the Cree, a double accommodation, but in neither case was the accommodation irrevocable. Within the Cree community one accommodation was between those who wished to continue their hunting way of life and those who wished to participate economically in the industrial development of their region. The intent to accommodate both visions was understood by negotiators as a main rationale for the agreement, and this was also clearly expressed by Quebec's main negotiator, John Ciaccia (1975), when it was first presented to the Quebec Legislature. A further accommodation had to be made between the Cree desire to maintain control over their own land – control that they had effectively retained during three hundred years of contact with Europeans – and Quebec's desire to exercise its jurisdiction over the area, which it had never previously done.

The wording of the initial clauses of the James Bay agreement concerning the extinguishment of aboriginal land title also might suggest that the Cree had taken an irrevocable step towards accepting industrial development. In voluntarily negotiating and signing the James Bay agreement with Quebec and Canada, the Cree had broken with some of their aboriginal political allies, notably the Indians of Quebec

Association (IQA), the organization to which the Cree belonged and under whose auspices the legal injunction against the La Grande Project had been launched. The action was successful in halting work, but the judgment was overturned after only one week. When this occurred the Cree were faced with a choice between taking their case to the Supreme Court of Canada or negotiating an out-of-court settlement. Their decision to negotiate was opposed by the IQA leadership, most vociferously by its head, Andrew Delisle, so that the Cree were forced to cut their ties with this organization. Subsequently, the IQA brought an unsuccessful legal action against the agreement itself. The Cree also incurred the strong disapproval of the National Indian Brotherhood (which later became the Assembly of First Nations), and specifically of the organization's high-profile leader of the day, George Manuel.

Given the optimism voiced by the Cree when they signed the agreement and their willingness to separate themselves politically from much of the rest of the Canadian aboriginal political movement, their current complaints regarding the implementation of the agreement and their opposition to the Great Whale project have been met with a degree of scepticism. Indeed, there are those in Quebec who assumed that the discontent and opposition of the Cree were merely stratagems in a negotiating position, designed to wring further concessions out of Hydro-Québec. Doubts have also been raised in Quebec over the arguments made by the Cree during the fight over Great Whale that they had negotiated the James Bay agreement "with a gun at their heads." In my view Cree opposition to Great Whale was not a negotiating strategy. Close examination of the Cree position when they were negotiating the agreement makes it clear that they were, in fact, under considerable pressure to get whatever deal they could and had few other options.

The Cree's prospects in 1972–74 for actually stopping the La Grande project were slight. They lacked sufficient funds for a prolonged legal fight. There was no legal requirement at the time, as there is today, for large development projects to pass an environmental and social impact assessment. Work on the project was proceeding twenty-four hours a day year round (apart from the brief halt due to the injunction and another, equally brief, due to labour problems), so that by 1974 it was clear that, even if the Crees were eventually to win a Supreme Court victory, it would come too late to stop the flooding, which was to begin in 1976. Moreover, a Supreme Court victory seemed unlikely; in 1974 the court had not yet shown itself unambiguously favourable to the principle of aboriginal rights. Finally, even if the Cree legal advisers thought they could actually win a Supreme Court judgment in time to stop the project, they knew the federal government could still pass legislation expropriating the land in question on the grounds that the project was vital to the national interest – an option the government might well have exercised, given the financial chaos that would ensue from a Supreme Court finding against the development.

Thus it is hardly surprising that by 1974 the Cree had concluded that they had no option but to negotiate. Supposedly they were negotiating a deal that would give Quebec "permission" to embark on a project; in reality the project was already half finished and was clearly going to be completed in any case. The Cree did, in effect, have "a gun at their heads." Their brief court victory had supported their conviction of the moral rightness of their case. Yet they also had to make the best of a bad situation and try to achieve whatever objectives they could through negotiation. They were certainly not happy to have had a hydroelectric project imposed on them. Cree "support" for the James Bay agreement was more a provisional acceptance, dependent on whether the implementation of the agreement met the specific objectives they had set. While this is not the place for a detailed analysis of the Cree experience of living under the agreement, it is fair to say that in significant areas the intended objectives are not being fulfilled.

The northern Quebec Inuit participated far more reluctantly than the Cree in the 1974–5 negotiations. Unlike the Cree, they had a significant minority group of dissidents among them who were opposed to the signing of the agreement, and Inuit leaders were less convinced that they were facing the only opportunity they would ever have to make a land claims deal. The La Grande project was being built on Cree territory, and its impact on Inuit territory was limited to the downstream effects of the diversion of the headwaters of the Caniapiscau River. In fact, the Inuit seriously considered pulling out of the talks, when Quebec presented a new and watered-down version of the environmental protection regime within days of the agreement's scheduled signing.

Today the Cree find themselves in a very different position from where they were in 1974. It has been their experience that many of the mechanisms of the agreement giving them rights over specific resources and influence over developments in the region have not functioned as intended. Nor have they been able to participate in the development of the region to the extent they had anticipated; a considerable proportion of young people remain without jobs. Their initial optimism and trust has been eroded as the Cree have had to resort to the courts to achieve even minimal government compliance from the governments of Quebec and Canada with many of the agreement's terms.

Meanwhile the legal and political powers of several other Canadian aboriginal groups peoples have become enhanced beyond even those won by the Cree and Inuit under the James Bay agreement. The Inuvialuit, Inuit of the Northwest Territories and some Dene groups have concluded land claims agreements that, in some cases, give them additional rights and powers, while groups that have never signed land claims agreements now have enhanced rights that flow from several recent Supreme Court decisions. Given that, from the Cree point of view, implementation of the agreement has fallen short of its intent and, in light of the gains made by other groups, it seems not to contain everything to which they were entitled in the first place, the Cree have tried increasingly to make other gains, at times by adopting legal and political strategies outside the framework of the Agreement.

Exclusionists and Accommodationists

Neither their disappointment with the implementation of the agreement nor their sense of having been bypassed by the legal and political achievements of other Canadian aboriginal groups fully explain why, in the early 1990s but not during the 1974–75 negotiations, the Cree so vociferously opposed the further industrial development of their territory. Nor do the greater opportunities offered to them by the Great Whale project, compared to the La Grande project, entirely explain why the Inuit were less reluctant to accept Great Whale in the 1990s than La Grande in 1975, or why it was they, and not the Cree, who sought an accommodation with the Great Whale project, despite their special interest in Manitunik Sound. In my view Cree caution towards the Great Whale project, as well as the greater Inuit openness to it, are not so much a matter of each group giving up its 1974 position as each returning to its own core cultural proclivities, or what Bourdieux (1977) refers to as "habitas."

In their relations with Euro-Canadians, the Inuit are generally inclined to take an "accommodationist" approach (or what McFeat has called the "extender" strategy), that is, to be open to whatever new opportunities Euro-Canadians may bring with them. It is the Cree who tend to take an "exclusionist" approach (McFeat's "restricter" strategy), that is, to be guarded in their relations with Euro-Canadians, for the most part incorporating outside innovations into their existing ways of doing things and maintaining social distance from outsiders.

(For a more detailed discussion of these concepts, see McFeat 1983; Tanner 1983.) These cultural proclivities are not set in stone. Accommodationist and exclusionist tendencies may coexist within a single group and may alternate occasionally as one or another internal faction gains political ascendancy.

In the case of ethnic groups in a multiethnic situation, I suggest that whether or not a group's behaviour is in accord with its own habitas will depend not only on the immediate circumstances of group self-interest but also on the behaviour of other groups with which it is in regular contact. Where several groups live in a structured relationship, one group's strategies, and the degree to which they conform to the group's own habitas, will be influenced by the strategies adopted by the other groups. An ethnic group under these circumstances may frequently find it advantageous to follow a strategy that is distinctive from that used by other such groups. It is thus not only in and of themselves but especially by comparison to the Inuit that I suggest that the Cree have general exclusionist tendencies.

A number of anthropologists who conducted research at Great Whale River in the 1950s and 1960s noted the same general tendencies in Cree and Inuit adaptation to outsiders (Balikci 1961; Barger 1977, 1980; Honnigmann 1952; Johnson 1962; Walker 1953; Wills 1969). During this period a radar base was built and put into operation at the site. Significantly, and in contrast to their view of the La Grande or Great Whale hydroelectric projects, the Cree and Inuit did not see the radar base as having a major impact on their hunting lands. Its effects were understood to be limited mainly to the area around the base adjacent to the two aboriginal settlements at Great Whale River; hence, on the surface its physical impact was the same for the Cree as for the Inuit. However, the social impacts for each group turned out to be quite different. The researchers noted that the accommodationist Inuit quickly took advantage of jobs offered by the base and promptly abandoned hunting to move into the settlement, making necessary adjustments to this new way of life. The exclusionist Cree, by contrast, continued for as long as they could to hunt and trap, did not seek paid employment, and kept their distance from the outsiders.

These differences in the way northern Algonquians and Inuit have reacted to outside contact have been seen elsewhere in northern Canada. In nineteenth-century Labrador, for example, the nomadic Inuit, having begun to trade with the new Moravian mission stations, soon began to settle in the communities the missionaries had established and quickly thereafter abandoned their shamans. In contrast, the Labrador Innu, an Algonquian group closely related to the Cree, retained their nomadic ways for at least another century while also making use of the new opportunities of trade. As with the Great Whale radar base, it was people of Inuit and part-Inuit descent who sought jobs at the Goose Bay Air Base when it was constructed in 1941. Very few Innu did so, even though the base was built within their traditional territory. Though they have now been settled in villages for several decades, the Innu have retained many of their animistic traditions. These same two kinds of opposing strategies indicate the differing basic cultural proclivities of the James Bay Cree and the northern Quebec Inuit.

During the James Bay negotiations the habitas of the Cree and the Inuit were temporarily inverted. The Cree were faced with the La Grande project, which was proceeding in any case, and they chose to embrace it in an attempt to minimize the damage and to participate in the inevitable development of their region. Equally uncharacteristically, the Inuit remained reluctant to endorse the development since it was taking place mainly outside their region and thus offered few opportunities for their participation. The signing of the James Bay agreement was not the only time that the normal approach of Cree and Inuit towards outsiders was reversed. During stable periods in the era of the Hudson's Bay Company fur trade monopoly, it was the Cree who became the accommodationists, especially one group that was known to the traders as the "home guard." The home guard lived close to the posts where they were often given employment, had close relations with the traders, and at times supplied the post with game meat. It was the Inuit who became, by default, the exclusionists. The reason for their inversion was that the fur trade was based on the demand for beaver, which inhabited Cree territory, the wooded interior where the Inuit seldom ventured. Beaver formed the basis of an economy that allowed the Cree to incorporate trapping within a way of life that consisted fundamentally of subsistence hunting (Francis and Morantz 1983). Under different economic circumstances, however, for example when beluga whales were hunted commercially for the Hudson's Bay Company, when white fox was in demand, or when a market for soapstone carving was established, then the Inuit were able to exploit their greater accommodationist proclivity with respect to outsiders, while the Cree retreated to their more typical exclusionist tendencies.

If we adopt this accommodationist versus exclusionist perspective, the question for this chapter becomes not so much why the Cree opposed the Great Whale project in the early 1990s as why they took the role of accommodationists during the James Bay agreement negotiations in 1974–75. Why did the Cree open themselves up to outsiders at that time and embrace a new way of life, as their apparent enthusiasm for the agreement suggests they did?

To the extent that the Cree became accommodationists in their negotiations over the Agreement, they did not fully abandon their exclusionist proclivities. As noted above, they made a compromise between the two positions, insisting on the inclusion of provisions to accommodate those who wanted to remain exclusionist – that is, to maintain the centrality of subsistence hunting in their way of life – while also seeking provisions that would allow them to participate in future development of their territory. By 1974–75 a whole generation of Cree had completed formal schooling, and these in particular were susceptible to warnings by influential outsiders – teachers, Indian agents, missionaries, even several anthropologists – that hunting was a dying occupation, and that it could not survive as a way of life. Fur prices had been in decline for years and many Hudson's Bay Company stores no longer gave credit, which full-time hunters had come to depend on. From this standpoint, the idea of industrial development as a source of future jobs seemed to make sense. Nevertheless, the Cree did manage to have a generous income security program for hunters included in the agreement.

But only a few years after the signing of the agreement, things began to look very different. A period of rapid inflation, particularly in the price of oil, had made it impossible for most Cree to continue to live in their settlements, eating store-bought groceries and heating their houses with oil. The kind of jobs that might have enabled them to sustain that way of life had not materialized in the Cree communities. Most of the available jobs involved living outside the community, which incurred a high risk of family breakdown. Thus, for many Cree subsistence hunting and fur trapping, supplemented by transfer payments, remained the best economic prospect, and the income security program turned out to be, from a Cree perspective, the most positive program in the agreement. For some Cree, in other words, the accommodationist strategy had failed.

In negotiating the James Bay agreement, the Cree were trying to make the best of a bad situation. They became accommodationists because they had to, not by choice. The Inuit, on the other hand, had a choice. Whether they were part of the agreement or not, they knew they would not be directly affected by the development for years to come and so might live to fight another day. Hence at the time of the negotiations it was useful for them to adopt the unaccustomed role of exclusionists.

The Development of Cree Ethnic Politics

At the time they signed the James Bay agreement the Cree were only just beginning to emerge from their erstwhile post-contact position as a politically powerless client group in relation, first, to the Hudson's Bay Company, more recently, to the federal government. They were thus a very different group from the Cree who, in the early 1990s, opposed the Great Whale project. Most adult Cree in the early 1970s were the clients, or permanent debtors, of the Hudson's Bay Company, an enterprise that was in decline, its power-broker role in the villages quickly being relinquished to various federal government agencies. Quebec had gone to the Supreme Court in 1936 in effect to opt out of jurisdiction over the Inuit, leaving the federal government responsible for the administration of all northern Quebec, there being no significant non-aboriginal population in the region for the province to administer. In the 1960s the province tried, with the formation of the Direction général du Nouveau-Québec, to reverse this situation by asserting a separate provincial administration of the region, but the attempt for the most part failed, leaving this unfinished business to be accomplished by the James Bay agreement.

At the same time, for most of the year the Hudson's Bay Company and both the federal and provincial governments were effectively powerless to exercise jurisdiction over the Cree and Inuit hunting camps and throughout the vast hunting areas. Cree hunters were self-governing, in full control of their own territory and in charge of their own means of production. In practical terms, they were as near to being sovereign as any state, and yet they were unaware that they were sovereign, since as yet they had no national or ethnic identity or ideology.

Quebec Cree ethnic consciousness, at least within the Canadian and international context, was only beginning to emerge in 1974–75. It was the announcement of the hydroelectric development project itself that had triggered the beginnings of this consciousness, bringing together Cree political leaders from all over the region for the first time. It was in the early 1970s, during the intense three-year period that went from the legal bid for an injunction against the La Grande project, through the agreement-in-principle, and finally to the negotiations leading up to the final agreement, that modern Cree leaders from every band broadened their political skills and developed the beginnings of their ethnic consciousness.

It took several years, however, for these leaders and for the Quebec Cree as a whole to develop and adopt an assertive ethnic identity. In my view, it is only in the past decade that the Cree have fully developed a consciousness of being Quebec Cree, and it is from this new ideological basis that they have been able to draw much of the political strength to pursue their independent objectives, including their opposition to the Great Whale project.

Conclusion: Cree Opposition to Great Whale

There are many reasons for the Cree turn-around from supporting the 1975 James Bay agreement to opposing the Great Whale project in the 1990s. Some of the reasons are dependent on external factors. For example, the standards now used in Canada and elsewhere for making public policy decisions about aboriginal matters and development projects have altered radically since 1975. The general practice of environmental impact assessment and the inclusion of protection regimes and mitigation procedures, pioneering as their limited inclusion in the agreement was, have now well outgrown their earlier incarnations. Such external developments have empowered the Cree beyond their situation in 1975.

Moreover, the issue of social impact, which is generally included (if somewhat uncomfortably) within environmental assessment, is proving to be a far more significant and contentious area of concern than was once realized. It is fraught with unresolved dilemmas, one of which is the difficulty of separating impacts specifically attributable to a project from changes that were already underway before the project began. This distinction between types of impact is of understandable importance to the proponents of development projects, given their concern with their own specific liabilities, the limits of which they want to establish clearly. However, both from the Cree viewpoint and from a public policy perspective, the distinction has a different kind of significance.

Suppose it could be objectively established, for example, that the Cree of a particular community were already experiencing symptoms of social breakdown prior to the start of a specific development project, due to the impact of recent urbanization, schooling, television, etc. For the proponents of development it could be important in the assessment process to establish that, once the project was built, they could not be held entirely responsible for all social pathologies and human tragedies that might occur in the community, since some of the causal factors were already in place.

Yet from a Cree perspective, it would be equally important to take into account, in deciding whether and under what limitations the project should be allowed to proceed, an evaluation of what the *com-*

bined effects of these two kinds of factors were likely to be. From this perspective, undertaking a development project should depend on whether these combined effects could create a socially unacceptable situation, even if the project itself could not be held solely responsible. Thus it is important to establish a public policy perspective that includes consideration of all the impacts, project related or not, one that is at arm's length from the project proponent, whose main concern, quite naturally, is delineating the extent of liability that can be directly associated with the project. An independent public policy perspective of this sort was, in my view, generally absent in the debate over Great Whale. And the awareness of the Cree, particularly those living in Whapmagoostui, that their concerns about cumulative social impacts were not being addressed only served to confirmed them in a position of outright opposition to the project.

Another factor influencing Cree attitudes is the greater general acknowledgment in Canada today, compared to 1975, of the idea of aboriginal rights, particularly over land and resources. Several of the land claims agreements that have been signed since the James Bay agreement allow aboriginal groups considerably more in the way of co-management rights over resources, while international conventions and guidelines on indigenous rights that have since been established also go beyond the level that the Cree obtained under the James Bay agreement. New Canadian Supreme Court jurisprudence, from Sparrow to Delgamuukw (Supreme Court of Canada 1990, 1997), indicates that the provinces have less power to override aboriginal use of natural resources than was assumed to be the case in 1975. As noted previously, agreements granting aboriginal self-government in excess of the powers obtained by the Cree in the James Bay agreement have now been concluded with some Canadian groups (e.g., the Nunavut agreement) or are being negotiated. While the James Bay agreement contains provisions for amendments, this assumes willingness on both sides to do so, and no such willingness has emerged. Given external developments in the area of aboriginal rights, the agreement, as long as it remain unamended, has effectively gone out of date.

As the Cree have argued, the agreement has not always been honoured, either in spirit, in that the Cree have not been able to participate significantly in the economic development of their region, or in the letter, in that they have found it necessary to obtain compliance through the courts. Disaffection with the way the agreement has been implemented provides an essential context for comprehending the emergence of Cree opposition to the Great Whale project. In the absence of clear demonstrations to the contrary, the Cree see the La Grande project as having had major negative social impacts on their communities, particularly those adjacent to it. And, in the end, in opposing the Great Whale project the Cree have reverted to their familiar strategy of exclusionism in their relations with outsiders.

Acknowledgments

First I want to thank Dartmouth College, and particularly Jean L. Hennessey, for inviting me to the conference where an early version of this chapter was first presented. I also thank Jim Hornig for his painstaking editorial work on this book, always conducted with great courtesy. Over the years my research among the aboriginal peoples of Quebec has benefited from many sources of financial support, including the Social Sciences and Humanities Research Council of Canada, the Grand Council of the Cree (of Quebec), the Naskapi Band of Schefferville, Quebec, and Memorial University of Newfoundland. I received helpful comments on earlier drafts from several people, and although not all of them are reflected in the end result, I would particularly like to thank Alan Penn of the Grand Council of the Cree (of Quebec), Colin Scott of McGill University, Pierre Senécal of Hydro-Québec, and Claire Gigantes of McGill-Queen's University Press.

REFERENCES

Balikci, Asen. 1961. "Relations inter-ethniques à la Grande Rivière de la Baleine, Baie d'Hudson." Ottawa: National Museum of Canada. *Bulletin* 173:*Contributions to Anthropology, 1959*, 64–107

Barger, W. K. 1977. 'Culture Change and Psychosocial Adjustment.' *American Ethnologist* 4, no. 3: 471–495.

– 1980. "Inuit and Cree Adaptation to Northern Colonialism." In *Political Organisation of Native North Americans*, edited by E. L. Schusky. Washington: University Press of America.

Bourdieux, P. 1977. *Outline of a Theory of Practice.* Translated by Richard Nice. Cambridge, UK: Cambridge University Press.

Canada. Royal Commission on Aboriginal Peoples. 1996. *Report of the Royal Commission on Aboriginal peoples.* 5 volumes. Ottawa: Royal Commission on Aboriginal peoples.

Ciaccia, J. 1975. "Philosophy of the Agreement." *James Bay and Northern Québec Agreement.* Quebec: Éditeur Officiel du Québec.

Francis, Daniel, and Toby Morantz. 1983. *Partners in Furs. A History of the Fur*

Trade in Eastern James Bay 1600–1870. Montreal and Kingston: McGill-Queen's University Press.

Goddard, J. 1991. *Last Stand of The Lubicon Cree.* Vancouver: Douglas and McIntyre.

Honnigmann, J.J. 1952. "Intercultural Relations at Great Whale River." *American Anthropologist* 54, no. 4:510–22.

Johnson, W.D. 1962. *An Exploratory Study of Ethnic Relations at Great Whale River.* Ottawa: Department of Northern Affairs and National Resources, Northern Co-ordination and Research Centre.

McFeat, Tom. 1983. "An affair to remember; Winoque 1965." *Culture* 3, no. 1:79–90.

Nakashima, D.J. 1990. *Application of Native Knowledge in EIA; Inuit, eiders and Hudson Bay Oil.* Ottawa: Canadian Environmental Research Council.

Napoleon, H. 1992. "Yuuyaraq. The Way of the Human Being." *Northern Notes,* no. 3 (May):1–35.

Niezen, Ronald. 1998. *Defending the Land. Sovereignty and Forest Life in James Bay Cree Society.* Needham Heights, MA: Allyn and Bacon.

Scott, Colin, and Kreg Ettenger. 1994. "Great Whale Environmental Assessment Community Consultation: Final Report for Wemindji and Eastmain." 2 parts. Report for Grand Council of the Crees (of Quebec)/Cree Regional Authority (under contract with Hydro-Québec).

Senécal, Pierre, and Dominique Égré. 1998. "Les impacts du complexe hydroélectique La Grande sur les communautés autochtones." *Recherches amérindiennes au québec* 28, no. 1: 89–104.

Shkilnyk, A.M. 1985. *A Poison Stronger Than Love: The Destruction of an Ojibwa Community.* New Haven, CT: Yale University Press.

Supreme Court of Canada. 1990. *Regina v. Sparrow* [1990] 1 S.C.R. 1075.

– 1997. *Delgamuukw v. British Columbia* [1997] 3 S.C.R. 1010.

Tanner, Adrian. 1979. *Bringing Home Animals. Religious Ideology and Mode of Production of the Mistassini Cree Hunters.* St John's: Institute of Social and Economic research.

– 1983. "Introduction: Canadian Indians and the Politics of Dependency." In *The Politics of Indianness: Case Studies in Native Ethnopolitics in Canada,* edited by A. Tanner. St John's: Institute of Social and Economic Research.

Usher, P.J. 1987. "Indigenous Management Systems and the Conservation of Wildlife." *Alternatives* 14, no. 1: 3–9.

Waldram, J.B. 1986. "Traditional Knowledge Systems: The Recognition of Indigenous History and Science." *Saskatchewan Indian Federated College Journal* 2, no. 2: 115–24.

– 1988. *As Long As The Rivers Shall Run; Hydroelectric Development and Native Communities in Western Canada.* Winnipeg: University of Manitoba Press.

Walker, W.B. 1953. "Acculturation of the Great Whale River Cree." MA thesis, Department of Anthropology, University of Arizona, Tucson.

Werner, O., and K.Y. Begishe. 1980. "Ethnoscience in Applied Anthropology." In *Indigenous Knowledge Systems and Development*, edited by D. Brokensha, D.M. Warren, and O. Werner. Lanham: University Press of America.

Wills, R.H.J. 1969. "Perceptions and Attitudes of the Montagnais-Naskapi of the Great Whale River Concerning the Western World." MA thesis, Department of Anthropology, University of North Carolina, Chapel Hill.

7 Contemporary Cree Art in Northern Quebec

A Northern Artist's Look at the Impact of James Bay Hydroelectric Development on the Art and Craft of the James Bay Cree

KESLER E. WOODWARD

Introduction

Since 1975 the Quebec-owned utility Hydro-Québec has been diverting and damming major rivers in the province to produce hydroelectric power on a very large scale. The damming of rivers and the introduction of roads, television, and other modern services have had a dramatic impact on the traditional way of life of the Cree Indians, the region's indigenous inhabitants. Many studies have been done and much has been said in political, scholarly, and popular forums about the positive and the negative aspects of these changes on the Cree. Employment patterns and income levels, life expectancy, educational opportunities, subsistence activity, and many other arenas have been examined in the light of the dramatic developments of the past twenty years. But little or nothing has been said, and almost no serious study has been done, on the way the art and craft of the James Bay Cree has been affected by those changes.

My interest in the effect of modern development on the artwork of northern native people is longstanding. In Alaska, I have worked in one way or another with the art and craft of the circumpolar North for the last eighteen years as a museum curator, art professor, and artist. As visual arts curator at the Alaska State Museum in the late 1970s and early 1980s, I travelled to Alaskan villages, talking with artists, looking at their work, and purchasing outstanding examples for the museum's collection. As a member of the art department at the University of Alaska since 1981, I have worked with native artists, taught courses on the art of the circumpolar North, and delivered papers on topics ranging from the changing imagery of northern native people to the possible impacts of climate change on the development of northern art.

While serving as a visiting research fellow at Dartmouth College from 1988–90, I became aware of the conflicts between large-scale hydroelectric development in the James Bay region and the Cree people there. I was curious to know how the effects of development – not just the physical changes to the landscape but the consequences of more modern modes of communication, increased funding for social and cultural projects, and other secondary impacts – were manifesting themselves in the work of Cree artists and craftspeople.

It had been my experience in Alaska that such development had both positive and negative effects. Obviously, the traditional culture is disrupted by the sudden influx of roads, rapid communications, the rise of a temporary cash economy, and other sweeping changes. At the same time, those very disruptions bring with them the possibility of more leisure time, more funding for artists' workshops and symposia, and other benefits that have potentially positive effects. Where, I wondered, did the balance lie between the positive and negative effects in northern Quebec?

In September of 1990 I had the opportunity, as part of a multidisciplinary group from Dartmouth College, to visit Quebec and observe some elements of the project. We spent a full day in Montreal being briefed on various aspects of James Bay hydroelectric development, and then we went north to Radisson and La Grande Two, the site of the largest Hydro-Québec dam and power generating complex (see Appendix B). There we talked with other officials and saw some of the installations firsthand. It was a fascinating trip, and I was powerfully affected by this first encounter with the land, people, and sweeping changes taking place in the James Bay region. On my return to the Dartmouth area, I began a series of large-scale paintings based on my response to the experience.*

While in Radisson, my interest in the impact of the large hydroelectric development on artists and craftspeople deepened. When I posed a question about the subject to a senior Hydro-Québec official involved in negotiation and liaison with the Cree, his response made it clear that he was unaware of any historical artistic tradition or contemporary artistic activity among the indigenous people of the region. It seemed to me then, and I have become increasingly convinced

---

*The resulting paintings were shown in a solo exhibition, Cree/Cri, at Alaska's Anchorage Museum of History and Art in February, 1991, and at the University of Alaska Anchorage Art Gallery almost a year later. For more on those exhibitions, see the brief catalogs produced for those exhibitions, (Soos 1991; Licka 1991).

Figure 2   Woman's hood, East Cree, James Bay, second half of the nineteenth century. Collection of Haffenreffer Museum of Anthropology, Brown University.

since, that this very lack of awareness might well have influenced the character of Hydro-Québec impacts on the Cree.

Determined to explore these issues further, I applied for and received a small research grant from the Quebec Ministry of International Affairs in early 1991. My intent was to visit some of the villages in the region, to see the art and craftwork firsthand, and to ask questions about what kinds of changes had gone on in recent years. I made it clear that I approached the subject not as an anthropologist or art historian but as a northern artist and exhibition curator with some personal experience of the effects of development on northern art – one who was interested in seeing how people in another part of the North had been affected by the same kinds of changes I had witnessed in Alaska in the past decade and half.

This summary of my impressions and tentative conclusions about the impacts of hydroelectric development in Quebec on the art of the Cree comes out of that perspective. Because I know of no other region in North America where so little is known about the native art and craft, I will begin with a brief look at the historical art and craft of the James Bay Cree.

The Historical Craft of the James Bay Cree

Though Cree art and craft is so little known, examples of Cree art from the James Bay region can be found in museum collections, and there are a few documents examining evidence of ongoing traditions up through the early 1970s. The historical materials are primarily of two kinds – elaborate nineteenth-century beadwork and somewhat simpler but traditionally "purer" ceremonial and utilitarian craft.

Among the most visually striking examples of James Bay Cree art are nineteenth-century beaded and woven articles of clothing. A hood from the collection of the Haffenreffer Museum of Anthropology at Brown University demonstrates the vitality of such pre-twentieth-century material. (The example here cited, along with a number of other similar objects from the James Bay area and neighboring regions, is more extensively discussed in Hail and Duncan 1990.) This type of hood, elaborately beaded in a floral pattern, is usually attributed to the James Bay Cree, and many variations were worn as early as the eighteenth century. According to anthropologist Kate Duncan (Hail and Duncan 1990), it represents the earliest datable use of the floral style in Cree work. Duncan notes that the same style appears in Cree floss embroidery of about the same period, though the beaded examples are more stylized, and that British whitework embroidery is probably the prototype.

Hoods were commonly worn by both men and women throughout the subarctic Canadian North. The type of hood depicted here was worn by women. The James Bay-style hood is notable for its extreme length – the back of the hood nearly touches the ground when the wearer is standing. An 1865 photograph shows James Bay women wearing such beaded hoods (Hail and Duncan 1990, 182). Little is known about when they were worn or what significance they had other than as decoration and protection.

It is important to note that this example probably comes from the Fort Albany area, on the western side of James Bay, rather than from the Quebec coast of the bay. I have been unable to find any examples of elaborate nineteenth-century beadwork from the Quebec coast in museum collections or literature. Either none were collected at the time and preserved, or such work may not have been undertaken in the villages on the eastern side of the bay (according to Canadian anthropologist Adrian Tanner, who has worked extensively with the Mistassini Cree of Quebec, the first option is likelier; personal communication, May 1993). The presence of this and other examples from nearby Fort Albany, however, makes it clear that there was a thriving tradition of ambitious, elaborate beadwork and embroidery among the James Bay Cree in that era.

Traditional James Bay Cree Material Culture

On the East side of James Bay, examples of the traditional material culture of the Cree are less spectacular but no less interesting. Extraordinarily little has been written about the traditional material culture of the eastern James Bay Cree. One of the few significant studies is Edward Rogers's *The Material Culture of the Mistassini*, written in 1967 but based on fieldwork done in 1953–54. What Rogers says about the Mistassini is true to a large extent of the eastern James Bay Cree as a whole:

The Mistassini, like other eastern subarctic Indians, are a people rather unique in some ways, when compared to most other North American Indians. The usual fate of the greatest number has been to become "reservation" Indians and to have their former way of life totally destroyed, with their material culture vanishing except as a fond memory or being modified in terms of native crafts. The Mistassini at the time this study was made had not experienced such a fate, in spite of the fact that they had been in contact with Euro-Canadians for approximately 300 years. It is expected, though, that they will experience it before too many years pass. What can account for this phenomenon? Their location deep within the eastern subarctic wilderness has in large measure shielded them from the pressures brought to bear on most other Indian groups in North America. Because of the particular environment in which this group found itself, the fur trader was the only outside agent to introduce significant changes during the long years of contact. And the fur trader desired that they maintain their former way of life, altering it only enough to exploit the fur-bearers as efficiently as possible. Neither the farmer nor the lumberman has had any interest in the country. Only recently has mining become important, and as it increases vast changes can be expected to occur in the material culture of the Mistassini Indians. (Rogers 1967, 1–2)

Rogers goes on to discuss the material culture of the Mistassini group in some detail, and the objects he describes, from the collections of the Royal Ontario Museum, are common to the James Bay Eastern Cree as a whole. Many of the objects remain the models on which craft forms are made for market by James Bay Cree artisans today.

Wooden ladles were used for taking grease from boiling foods and sometimes as scoops for flour when making bannock, a type of biscuit. Rogers notes that the implement appears to have had greater ceremonial than utilitarian significance. At every feast, a ladle was prominently displayed but used only to deposit an offering of food in the fire or for drinking a little grease. Traditional grease ladles, made from birch, often had a geometric design incised on the handle, and the ladle was dyed, yellow to red-brown, with the bark of alder or willow. Fish spoons, also made of birchwood, were used for cooking and serving boiled fish.

Birchbark containers of all kinds were used traditionally, at least in areas where birch trees grow. A typical James Bay Cree birchbark container had a square bottom, rising to an oval rim higher at the ends than the sides, with a separate root rim attached. Skin bags were employed as well, especially in the northern portions of the region where there are no sizeable birch trees. Cloth bags, on the other hand, were introduced to the Cree. Small bags decorated with embroidery were made to hold hymnals and prayer books. Larger bags were used to hold a drum. Cloth also replaced skin in the large sacs used to hold bedding while travelling, and small roll-up bags came to be used by women for storing needles and other sewing equipment.

Shot pouches for carrying ammunition, similar to nineteenth-century fire-bags for carrying flint, steel, and tinder, were also used by the Mistassini Cree and can be found in the collections of the Royal Ontario Museum and the National Museum of Canada.

Five styles of snowshoe were traditionally used by the James Bay Cree – elbow, bear-paw, swallow-tail, beaver-tail, and pointed. The frame is made from green birch sapwood. They were made usually in late fall or early winter. The thongs were made from moose or caribou hide. Tassels were sometimes placed at the centre of the toe and along the sides of the tail of the frames. While the snowshoes were generally made by men, lacings were mainly made by women from unsmoked caribou and moose hides.

Among the musical instruments of the Cree – drums, rattles, violins, guitars, and mouth harps – the drum is significant in that it is employed for religious purposes. The traditional drum was circular, with two heads. Its body consisted of a piece of larch approximately five feet long and three to four inches wide, bent into a cylinder. Its ends were held together with nails and thong. Each head was made of one caribou hide scraped very thin. The suspension cord attached to the lacings was of moosehide or string. A fine thread snare was stretched between the outer hoops, across the heads, and four or five pieces of wood were attached to each snare. Sometimes the drums were decorated, with a circle of red dots about the circumference of each head. At the centre of each head was a small circle of red dots through which passed a cross of red dots. On occasion, a realistic design was painted in place of the circle and cross. According to Rogers (1967), these designs may have depicted a successful hunt.

Like other unacculturated hunting-gathering groups, the Cree made these traditional objects for both ceremonial and practical use in elegant traditional forms that displayed excellent craftsmanship. While the objects were often decorated, they were not thought of as "art" by their makers. What is most significant for our present purpose is that the objects being made by the James Bay Cree as late as the 1950s were substantially identical to artifacts from the nineteenth century and even earlier. Moreover, according to Canadian anthropologist Adrian Tanner (personal communication, May 1993), virtually all of the types of objects described by Rogers in 1967 are still being produced today, for personal use and as objects for market as art or craft.

Marketing Cree Arts and Crafts

The move towards marketing objects based on these traditional forms began in the 1960s. Well-known anthropologist Nelson Graburn, who did extensive fieldwork in the eastern Canadian Arctic and Subarctic as early as 1953, examined an early Cree marketing effort in 1971 (Graburn 1978). He notes that beginning in the late 1960s, what came to be known as Cree Craft was promoted by the Fédération des Coopératives du Nouveau-Québec. The Cree had unsuccessfully tried several commercial arts and crafts forms in previous years – first, copying Inuit soapstone carving (which the Canadian government stopped), then making soft-stone "dolomite" animals and figurines, then wooden figurines, and finally, wooden replicas of their traditional implements. Federation employee Andrée Roegier worked with the Cree of Great Whale River in developing Cree Craft, ensuring that the forms, colours, and motifs were authenticated by elders.

Graburn (1978) brought examples of Cree Craft to exhibit at the Berkeley campus of the University of California in 1971. Produced for commercial markets since 1969, the crafts were mostly full-size copies of traditional implements and made almost entirely of wood. They bear a striking resemblance to the traditional objects described by

Rogers. While the Cree Craft in the exhibit were made by numerous men in Great Whale River and Wemindji, the designs were applied in bright acrylic paints by two or three men in a local cooperative at Great Whale River.

Graburn describes an experiment he conducted at Berkeley in which he interviewed visitors to this exhibit and to a similar exhibit of work by Inuit soapstone carvers. He asked interviewees for their aesthetic opinions of the work. The title of Graburn's study (ibid.) is taken from the negative response of one viewer to the Cree Craft – "'I Like Things to Look More Different than That Stuff Did': An Experiment in Cross-Cultural Art Appreciation." The study illuminates the various obstacles to the development of Cree art, many of which have proved difficult to overcome in the past and remain intact today.

1. What do you think of this? (Answer) Well, what particular feature or pieces are you reacting to in making these judgements?
2. What pieces do you like best and/or worst, and why?
3. In these pieces what do you think the creators were trying to get across, or at least what do you get out of them?
4. What other kinds of arts and crafts do you like or know or have at home/experience. (Graburn 1978, 54)

About sixty percent of viewers responded negatively to the Cree Craft. Positive responses focused on simplicity, balance, symmetry, good shapes and colours, and overall design, in that order. But negative responses identified plainness, oversimplicity, and poor choice of colour. One frequent comment was that the painted designs and their central placement worked against the basic wooden forms (ibid., 55–6).

Graburn reports that viewers who responded positively admired the workmanship, commenting on the smoothness and regularity of the forms. But the excellent craftsmanship of the Cree worked against them as well, and an overwhelmingly negative series of responses commented that the objects looked almost mass-produced, too well finished, too new, too clean, cheap, and plastic. Several viewers said they expected Indian crafts to look "primitive, rough, subdued, or at least old." One person said, "It would be in the Cree's interest to dirty them up a bit before exhibiting them," and many expected them to be worn, weatherbeaten, used looking. People were concerned about "authenticity," and most preferred the snowshoe, which looked "authentic," to the crafts. A number of viewers changed became more negative in their responses after learning that the objects were made for sale. One young woman said,

"I thought it was disgusting that they had to paint them in order to make them saleable" (ibid., 57).

The reaction to the Canadian Inuit carvings, much better known than "Cree Craft" was strikingly different: ninety-eight percent of viewers responded positively. Positive responses primarily conveyed admiration for the traditional Inuit way of life and their closeness to nature and the animal world, so clearly communicated in their artwork. While viewers of the Cree Craft seemed to think that having to produce crafts was a form of oppression for the Cree, the assumption with the Inuit carvings was that the artists were happy and fulfilled. There was little concern with authenticity, though the text accompanying the carving exhibition made it clear that this was a commercial art form introduced to the North in 1949 and no more "traditional" an activity than the production of Cree Craft (ibid., 58, 63–4).

Graburn speculated that non-native expectations regarding Indian arts and crafts are based on public familiarity with Northwest Coast and Plains Indian objects in museums; hence the notion that all Indian-made objects should be old, rough, made from dark wood, and coloured, if at all, only in earth tones. Never having seen any precommercial Inuit art or handicrafts, viewers had no preconceived ideas, and the images of men and animals seemed to bear out the romantic image of the Inuit we are steeped in from childhood.

The viewers were also concerned with categories of art forms (ibid., 64). The Inuit work was illustrative, or evocative, while the Cree handicrafts were utilitarian, or at least reproductions of utilitarian objects. Cree Craft posed the dilemma whether it should be judged as "functionally adequate" or "formally decorative." The Inuit carving presented no such dilemma.

Finally, Graburn points out that the audience drew great satisfaction from the easily understood "messages" of the Inuit carvings, while the Cree Craft was seen as too utilitarian in form and the decoration "unreadable." In the end, the Cree objects were "replicas," despite being authentic and true to traditional forms. The Inuit sculptures were not replicas of anything traditional but original ("authentic") versions of what they appeared to be – models of exotic and interesting aspects of Inuit life (ibid., 66–7).

Thus, the early marketing effort of the Cree was a commercial failure, undoubtedly for the reasons indicated by Graburn's study. By visiting Cree villages and Quebec marketers, I hoped to discover what has taken its place, and whether the character of the new work has been affected by the outside influences brought into the villages by Hydro-Québec development.

Cree Art and Craft in Northern Quebec Today

Since so little attention has been paid to the art of the Cree, it was not until I reached the outskirts of the James Bay region and began to meet people working for or with the Cree that I started to get any real information on what kind of arts and crafts were currently being produced in the area. These people are acknowledged and thanked at the end of the chapter.

What follows here is a summary of the kind of work I saw being made for display and sale in the region, and the tentative conclusions I was able to draw about the impacts of large-scale hydroelectric development on the forms and development of the art.

Sculptural forms fashioned from tamarack branches in the form of traditional goose hunting decoys were among the most common craft items in evidence throughout the region – not just in the craft stores but in piles on shelves in variety stores. They varied in size and quality of craftsmanship, from the elegant productions of John Blueboy of Waskaganish, the originator of the current form, to less-ambitious, hastily produced examples by unnamed artisans. Blueboy, born in 1930, is said to have revived this vanishing craft at age thirty-five, during his convalescence from tuberculosis, and most of the decoy forms still come from his community.

Many soapstone (steatite) carvings were displayed in each of the craft stores. The carvings ranged from small, roughly finished likenesses of fish and other animals strung on rawhide thongs as pendants, to elaborate, more elegant, larger forms of birds and multiple animals executed as freestanding sculptures. I was told that virtually all of the carving is done by men, that almost all of it is done in Great Whale, and that the stone itself comes from the Belcher Islands. Steven Sheshamush, born about a hundred miles north of Great Whale in 1912, is the best-known Cree carver. A trapper in early adulthood, he began carving soapstone about 1960, when his trapping activities had slowed down. He has been invited to carving festivals in Europe, Scandinavia, and Japan. Many of the soapstone carvings I saw were by Sheshamush or other members of his family. There were also wooden carvings in the shape of animals, again often made by members of the Sheshamush family, but much of the work in wood took the form of reproductions of such traditional items as grease ladles, fish spoons, and the wooden portions of snowshoe models, rattles, and drums. These forms were remarkably similar both to the traditional forms described by Rogers (1967) and others, as well as to the Cree Craft of the early 1970s described by Graburn (1978).

Various articles of clothing are based to one degree or another on

Figure 3   Wachiya craft store – Val d'Or.

traditional patterns and decorated in both traditional and non-traditional ways. Skin sewing and beadwork were prominently featured in each of the craft stores. Many of the bags, mukluks (skin boots), and other embroidered or beaded clothing items are clearly based both in form and decoration on historical models. (See, for instance, Burnham 1992; Hail and Duncan 1990; Noel 1979; and Rogers 1967, for examples and discussion of historical models.) Most beaded or embroidered items, however, took the simpler form of circular pendants for necklaces and earrings. Many of the beaded necklaces incorporated designs characteristic of Indians in the southwestern United States and elsewhere. The makers of skin objects and beadwork were rarely identified. Beads and other craft items are sold to artisans in all of the craft outlets, as are how-to books on beadworking and other crafts. A particularly extensive stock of craft materials was available at the Cree-owned art and craft store at Wachiya, in Val D'Or. This impressive store/gallery contains by far the largest single collection of contemporary James Bay Cree art and craft to be found anywhere.

Beautifully made, full-size snowshoes of traditional design, notably those produced by Emma and Johnny Shecapio and their family in Mistissini, were for sale in each of the craft stores. Johnny Shecapio crafts the frames while Emma selects and prepares the laces from moose and caribou hide. The geometric patterns on the snowshoes

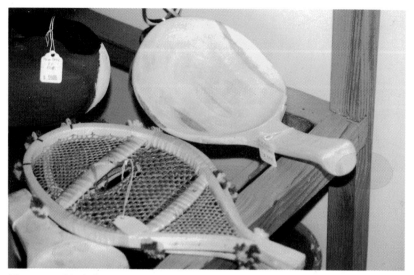

Figure 4   Model snowshoe and grease ladle.

are traditional, as are the patterns of the red woolen pieces that are attached. According to information provided by Wachiya, with their ten children the Shecapios produce about twenty-four pairs of snowshoes per year.

In addition to these crafts, two-dimensional artwork in a variety of media was featured at Wachiya in Val D'Or and the craft store in Chisasibi. (This store, much smaller than the store at Wachiya and in a smaller community, does almost as brisk a business. Surprisingly, its customers are almost exclusively Cree members of Chisasibi itself. Such extensive sales of native artwork to members of the same native community is unusual.) Painting and printmaking both seem to be growing in prominence. Accomplished and innovative mixed-media works draw upon the local landscape, animals, people, and spiritual symbols as sources of imagery. Among the most acclaimed regional figures are Virginia Bordeleau of Val D'Or, Glenna Matoush of Mistassini, Morley Stuart of Wemindji, and Tim Whiskeychan and Allan Linkletter of Waskaganish. (For more on these and other painters and printmakers of the region, see Roussan and Ruph 1990. Matoush's work is also reproduced and briefly discussed in Leblond 1989.) Glenna Matoush's two-dimensional work is probably the best known in the region. Born on an Ojibway reserve in Ontario, Matoush began studying painting in 1964. She settled in Quebec in 1970 and has devoted herself primarily to etchings.

Figure 5   Craft store – Great Whale River.

Observations, Tentative Conclusions, and
Many More Questions

I was able to see a broad and representative selection of contemporary
Cree art and craft in the Cree Council-owned art and craft stores in Val
D'Or, Great Whale River, and Chisasibi. Time and lack of funding precluded visits to the smaller communities I had originally hoped to
explore, but while the visits would have been interesting, they would
have shed little additional light on the questions I was asking since the

work produced in those villages (all of which have their own small craft stores) is well represented in the three larger craft stores.

Overall, I took more than one hundred slides of contemporary Cree art and craft. I also got a lot of answers to specific questions about how the craft stores operate; how, when, and where the objects are made and by whom; and what kinds of changes have taken place in recent years. I saw substantial amounts of work by virtually all the regionally prominent artists and craftspersons, and by many more unnamed artisans. A few questions, answers, and observations will give some idea of the kind of information I sought and found.

I asked why the makers of the arts and crafts were so rarely identified; I was told that, at least at Wachiya, the largest craft store, the managers had formerly put names to the handicrafts but that those who were less accomplished than others had asked that their names be removed. I asked whether they would give names if the makers asked them to; they would, I was told, but the makers would not ask; they were "too shy," especially the women. Objects made by men are far more often identified by name.

I asked whether the items were purchased or consigned and was told that everything was bought outright, either from the artists or from the smaller village stores. Wachiya does not own the other outlets, but Wachiya's managers work for the Cree Council (as do the managers of the smaller stores) and their job is not just to run the store but to provide technical assistance, training, etc., to artists and to the managers of the village craft stores. Craftspeople tend to produce more work when they need money and are thus eager to be paid for it at the time and less willing to wait for payment on an often lengthy consignment. I was also told that the trappers' cheques (from the subsidy given under the James Bay agreement of 1975 to hunters and trappers who spend time in the bush in traditional pursuits) come in four times a year and are often used to fund trips into town, where the people hope to sell their crafts.

When I asked how much art and craft was being produced now relative to five or ten years ago, I was told, somewhat to my surprise, that it was about the same – that there had always been artisans among the people, and that before the inception of the Cree Development Corporation, the Cree Council subsidiary that funds the craft stores, craftspeople did the same kind of work while out in bush camps, when there was extra time, and sold it at the local Hudson's Bay (now Northern) store.

I also asked about quality. Had it improved in recent years? The answer was an emphatic, unanimous "yes." The craft stores exercise more overt quality control than they did ten or even five years ago, refusing to purchase work of poor quality.

Finally, I asked about the future of marketing and sales outlets for Cree arts and crafts. The quality and quantity of what I saw in just three communities led me to believe that there must be similar stores in Montreal, Ottawa, and elsewhere, and that they could be highly successful and might encourage greater production and even more ambitious work by the artists. It was patiently explained to me that there is simply not enough supply, and not a big enough population base to increase the supply substantially.

This sobering reminder of their small population brought home the slim prospects for the continuing growth of Cree art, at least in terms of visibility and marketability. But it also showed how remarkable the achievement of Cree artists and craftspeople was. Out of a population altogether less than that of a single small town in the more urban regions of North America, I had seen outstanding work produced in a wide range of media by scores of craftspeople, and intriguing, even more ambitious work by as many as a dozen artists who are known and admired throughout the region and, in some cases, beyond.

What, then, is the effect of large-scale hydroelectric development on the art of the James Bay Cree? Many questions remain before that central issue can be answered with any degree of certainty. The role of non-market arts and crafts and the use of traditional objects by people in the communities and in the bush should be explored. The role of the public schools in helping to sustain traditional arts and crafts forms should be examined. Individual artists and artisans, not just those who market their work, should be interviewed for their impressions of changes. These areas and others remain virtually untouched, and I know of no one who is proposing to make a study of the subject.

Nonetheless, it is possible to say some things about the effect of recent development in the region. Not surprisingly, it has brought both curses and blessings. That the culture has been substantially disrupted is indisputable, and the loss of traditional ways and means to what was not so long ago one of the last relatively unacculturated hunting-gathering groups in North America has to undercut the traditional forms and values on which the art is based. The daily connection of the people to their traditional forms is less clear as the character of their lives departs increasingly from those traditional ways.

At the same time, the influx of money that has come with the various agreements and settlements and the greater access and communication made possible by changes in the region have undoubtedly opened new avenues of expression as well as new outlets for ambitious artists and craftspeople. Now run by the development corporation, the craft stores serve as effective outlets, training centres, and sources of supply for materials from which arts and crafts are made. Moreover, workshops and symposiums are beginning to take place in the region, and artists from different villages are getting together with one another and with artists from outside the region to discuss mutual concerns and share ideas and enthusiasms. Virtually everyone with whom I talked was excited about the possibilities for growth in range of expression and collaboration in the development of new ideas.

But even these positive developments in the scale and effectiveness of marketing, access to new materials and ideas, and awareness of a broader range of formal and conceptual possibilities have a potential downside: the blessings that help balance the curses raise questions of their own. Some of the questions are easy. Surely a painting or print done by a Cree artist that renders a traditional Cree theme in a modernist mainstream style or one made in response to a changing way of life, is as "authentically Cree" as more traditional forms. And surely the Cree, like other native North Americans, did not survive for millennia in a hostile environment by ignoring the technological and material innovations that came along.

But other questions are harder to answer. When a Cree artisan uses modern materials to create beadwork in patterns drawn from a book of southwestern United States Indian designs, to what extent is the Cree beadworking tradition being preserved and fostered? Is it enough that the Cree tradition of careful, elaborate decoration of personal items is being encouraged, not only through better access to materials and better outlets for the sale of such work but through increased leisure time in which to pursue such tasks? Further, to what extent are the designs and materials themselves being dictated by market forces, rather than by personal artistic decisions or cultural predilections?

These and similar questions are not unique to the Cree or to the situation in the James Bay region. They confront both the artists who make and all the others who sell, buy, study, exhibit, or simply appreciate the work of native people who are undergoing rapid acculturation in various locales around the globe. Though they are interesting questions, they are not, finally, about the balance of good and ill in the changes being wrought by large-scale hydroelectric development in Quebec but about what the term "native art" can possibly mean in a modern world. As anthropologists and art historians address that issue, the responses of Cree artisans to the startling changes of the past generation provide not only another case study but further testimony to the resilience and adaptability of such cultures.

It would take much more than a nine-day trip to James Bay to reach any kind of conclusion as to where the balance in all these changes lies – whether the net is gain or loss. What is clear is that there are thriving arts and crafts traditions among the Cree of James Bay, just as there have been for countless generations. The nature of those traditions and the motivations, character, and style of what is being made now are changing more rapidly today than a century or even a few decades ago. But this is a living, thriving tradition of which the Cree people are justly proud, and it should be more widely acknowledged, better studied, and more fully appreciated both within the region and in the wide world outside James Bay.

Acknowledgments

What I saw, the questions I asked, and even where I went were very strongly influenced by information and suggestions I received from various people during my brief stay in Quebec. Through Alan Penn (project manager for Environment and Lands for the Cree Regional Authority) I had the good fortune to meet Bill Kemp and Lorraine Brooke in Montreal. Kemp, a cultural geographer and former professor at McGill University, was at the time engaged in contract work producing cultural maps for the Cree. Lorraine Brooke is a geographer who was then engaged in similar work for the Cree. These two widely respected figures were very helpful in providing an initial overview of what I might expect to find in the way of contemporary Cree art/craft and made very helpful suggestions about others whom I might contact.

In Val D'Or, I met Susan Marshall, an anthropologist employed by the Grand Council of the Cree, and David Denton, an archaeologist who was at the time doing contract work for the Cree Council, excavating sites in Great Whale River (Whapmagoostui) and making cultural site maps of the communities in cooperation with the people in the villages. Both provided specific names of artists and craftspeople whose work I might see, helped to educate me about recent developments in Cree art, and, like everyone with whom I spoke, pointed me to Great Whale River and Chisasibi as the key communities to visit.

Muriel Buteau, the assistant manager of the Cree Council-owned art and craft store, Wachiya, in Val D'Or, allowed me to photograph the work in the store and ask questions about the artists, their work, the operation of the store, and state of Cree art and craft in general. This was perhaps the most important meeting I had in terms of information on specific kinds of art and craft, specific artists, and what I might find in the smaller communities. I am grateful to all of these people, and to many others I met during my brief stay, for their kindness, openness, and willingness to share their insights with an unexpected stranger from a distant but in many ways similar part of the North.

REFERENCES

Biersteker, Thomas R. 1993. "Constructing Historical Counterfactuals to Assess the Consequences of International Regimes." In *Regime Theory and International Relations*, edited by V. Rittberger. Oxford: Clarendon Press.

Bourassa, Robert. 1985. *Power from the North*. Scarborough: Prentice-Hall.

Burnham, Dorothy K. 1992. *To Please the Caribou: Painted Caribou-Skin Coats Worn by the Naskapi, Montagnais, and Cree Hunters of the Quebec-Labrador Peninsula*. Toronto: Royal Ontario Museum.

Canadian Arctic Resources Committee. 1993. "Creating Nunavut and Breaking the Mould of the Past." *Northern Perspectives*, 21, no. 3:1–18.

Freeman, Milton M.R. 1989. "Graphs and Gaffs: A Cautionary Tale in the Common-Property Resources debate." In *Common Property Resources: Ecology and Community-Based Sustainable Development*, edited by F. Berkes. London: Belhaven Press.

Freeman, Milton M.R., and Ludwig N. Carbyn, eds. 1988. *Traditional Knowledge and Renewable Resource Management in Northern Regions*. Edmonton: Boreal Institute of Northern Studies.

Goudie, Andrew. 1994. *The Human Impact on the Natural Environment*. 4th ed. Cambridge, MA: MIT Press.

Graburn, Nelson H.H. 1978. "'I Like Things to Look More Different than That Stuff Did': An Experiment in Cross-Cultural Appreciation." In *Art in Society: Studies in Style, Culture, and Aesthetics*, edited by M. Greenhalgh and V. Megaw, 51–70. New York: Saint Martin's Press.

Hail, Barbara, and Kate Duncan. 1990. *Out of the North: The Subarctic Collection of the Haffenreffer Museum of Anthropology*. Providence, RI: Haffenreffer Museum.

Leblond, Jean-Claude. 1989. "Un Contexte Culturel Différent. *Vie des Arts*, 34, no. 137 (December): 32–35.

Licka, C.E. 1991. *Inscapes and Poethics: Speculations on the James Bay Paintings*. Anchorage: University of Alaska Anchorage Art Gallery.

Noël, Michel. 1979. *Art Décoratif et Vestimentaire des Amérindiens du Québec*. Ottawa: Les Éditions Leméac.

Rogers, Edward S. 1967. *The Material Culture of the Mistassini*. Edited by A.S.N. 80. Ottawa: National Museum of Canada Bulletin.

Roussan, Jacques, and François Ruph. 1990. *Abitimi-Temiscamingue in Painting*. Sainte-Claire, Quebec: Roussan Editeur.

Soos, Frank. *Cree/Cri: James Bay Paintings by Kesler Woodward*. Exhibition catalogue. Anchorage, AK: Anchorage Museum of History and Art.

Chronology of Hydro-Québec Development on James Bay

30 April 1971	Premier Robert Bourassa announces plans to harness the hydroelectric potential of the James Bay basin.
December 1971	The Société de'énergie de la Baie James (SEBJ) is established for the stated purpose of harnessing the hydroelectric potential of the rivers on the Quebec side of James Bay.
Fall 1971	Construction of road between Matagami and the site of La Grande 2 begins.
January 1973	Commencement of work on La Grande 2, first construction site in the La Grande River complex.
15 November 1973	Judge Albert Malouf grants an interlocutory injunction. This had been sought eleven months earlier. Work is suspended.
22 November 1973	Quebec Court of Appeal overturns the injunction
29 November 1973	The Quebec government submits settlement proposals, marking the start of lengthy negotiations.
August 1974	The Grand Council of the Cree (Quebec) is formed, bringing the (then) eight Cree communities of the James Bay region under one political structure. Billy Diamond becomes first grand chief.
November 1975	Signing of the James Bay and Northern Quebec Agreement (JBNQA).
1977	La Grande complex Environmental Monitoring Network is established.
1 October 1978	SEBJ becomes a wholly owned subsidiary of Hydro-Québec.
27 October 1979	La Grande 2 is commissioned.

December 1981	La Grande 2 generating station in full operation.
September 1984	Approximately ten thousand caribou drown while crossing the Caniapiscau River. Hydro-Québec denies that the incident was caused by releasing water upstream. Environmental organizations disagree.
1986	Establishment of the James Bay Mercury Committee.
1989	Maine terminates a $15-billion contract with Hydro-Québec for imported electric power
January 1992	A memorandum of understanding regarding preparation of impact statement for the Great Whale Project is signed.
March 1992	Governor Mario Cuomo cancels New York's $12.5-billion contract for power from Hydro-Québec, stating that it is no longer economically advantageous.
January 1993	Hydro-Québec and the Cree sign the Opimiscow agreement to finish construction of the La Grande River project.
August 1993	Hydro-Québec releases its environmental impact study on the Grande Baleine project.
November 1994	Quebec premier Jacques Parizeau announces that the Grande Baleine phase of James Bay development will be postponed indefinitely.
March 1995	Tufts University divests itself of its Hydro-Québec bonds.
August 1998	The Canadian Supreme Court rules that Quebec does not have a right to separate from Canada under international law but provides language the Cree find favourable to their option for separation.
November 1998	Hydro-Québec proposes a diversion of the Rupert River into the La Grande generating system and offers the Cree the opportunity to jointly invest in the project.

The Authors

RAYMOND COPPINGER is a professor of Biology at Hampshire College in Amherst, Massachusetts. He started exploring the great northern forest in 1971 when he and ten ecology students from Hampshire College canoed three hundred miles down the Albany River into James Bay. He has published papers on wolf behaviour in northern Minnesota and consulted with paper companies as far west as Vancouver. He and his wife Lorna have supplied trained sheep-guarding dogs to farmers in British Columbia. With economist Stan Warner, Coppinger makes annual summer trips to Canada, studying the paper and electic power industries.

JAMES F. HORNIG is the Dartmouth Professor of Chemistry and Environmental Studies, emeritus, at Dartmouth College in Hanover, New Hampshire. He was cofounder of the interdisciplinary undergraduate Environmental Studies program at Dartmouth in 1970 and chaired it until his recent retirement. He first saw Hudson Bay on a canoe trip down the Winishk River and has since explored other northern rivers by canoe, including the Coppermine and the Noatak.

BILL D. ROEBUCK is professor of Toxicology at Dartmouth Medical School where his teaching and research interests are at the interface of problems associated with diet and nutritional status in the presence of environmentally available toxic chemicals. He is an adjunct professor in Environmental Studies at Dartmouth College where he teaches a course on environmental health, which includes a major case study of mercury and the James Bay development. He has spent much time in the North, sea kayaking along the Quebec coast and in the Belcher Islands of Hudson Bay.

WILL RYAN is a nature and outdoors writer who teaches expository writing at Hampshire College in Amherst, Massachusetts. His waterfowl articles have appeared in *Outdoor Life, Sports Afield, The Shooting Sportsman, The Conservationist,* and elsewhere. He has a special interest in the declining black duck and Canada goose populations in the Atlantic flyway. He has written a book on smallmouth bass and is completing another on northern pike.

ADRIAN TANNER is professor of Anthropology at Memorial University, St John's, Newfoundland, where he has taught since 1972. He has conducted research on aboriginal people in various parts of northern Canada, published a book on the hunting rituals of the Cree Indians and another on the development of ethnic politics among Canadian Indians. He has also conducted research on various practical issues, including aboriginal land claims, social impact assessment, and expert court testimony. He headed a recent study of aboriginal governance in Newfoundland and Labrador for the Canadian Royal Commission on Aboriginal Peoples.

STANLEY WARNER is an Economics professor at Hampshire College in Amherst, Massachusetts. His teaching and research are centred in ecological economics and political economy, specifically economic development, population growth, and social change in local communities. He travels regularly to Ontario and Quebec with Ray Coppinger, reserching forestry and hydroelectric development and the relation of those industries to Cree life. He and Ray Coppinger fancy themselves as experts on fishing in the vast Canadian rivers and reservoirs.

KESLER E. WOODWARD is painter, curator, and art historian. He is currently professor of Art at the University of Alaska Fairbanks, where he has taught since 1981. Prior to coming to the University of Alaska, he served as artistic director of the Visual Art Center of Alaska and Curator of Art at the Alaska State Museum. From 1988–91, while on leave from the university, he was a visiting research fellow with the Dickey Endowment for International Understanding at Dartmouth College. He is the author of four books on Alaskan Art, most recently among them *Spirit of the North: The Art of Eustace Paul Ziegler* (1988), published by the Morris Museum of Art. Woodward's paintings are included in all major public art collections in Alaska and in public, corporate, and private collections on both coasts of the United States. Juried and invitational exhibitions that include his work range from Alaska to Brazil and Russia.

ORAN R. YOUNG is professor of Environmental Studies and director of the Institute of Arctic Studies at Dartmouth College. Long interested in the political economy of nature and natural resources, he discovered the Arctic and the Subarctic in the 1970s as areas offering unlimited opportunities to engage in relevant fieldwork, and he has been going north on a regular basis ever since. Today, he conducts or supervises projects involving Alaska, Canada, Greenland, Fennoscandia, and the Russian North. He and his wife, Gail Osherenko, an environmental lawyer who works in Siberia, have travelled far and wide throughout the circumpolar North.

Index